GIS-Based Studies in the Humanities and Social Sciences

GIS-Based Studies in the Humanities and Social Sciences

Edited by
Atsuyuki Okabe

CRC Press
Taylor & Francis Group
Boca Raton London New York

CRC Press is an imprint of the
Taylor & Francis Group, an **informa** business

A TAYLOR & FRANCIS BOOK

CRC Press
Taylor & Francis Group
6000 Broken Sound Parkway NW, Suite 300
Boca Raton, FL 33487-2742

First issued in paperback 2019

ISBN-13: 978-0-8493-2713-1 (hbk)
ISBN-13: 978-0-367-39195-9 (pbk)

Library of Congress Card Number 2005048572

Library of Congress Cataloging-in-Publication Data

GIS-based studies in the humanities and social sciences / editor, Atsuyuki Okabe.
    p. cm.
    Results from a six year research project entitled Spatial Science for the Humanities and
    Social Sciences (SISforHSS) carried out June 1998 to March 2004 by the Center for Spatial
    Information Science (CSIS) at the University of Tokyo.
    Applies spatial methods in particular to economics, human geography, and archaeology.
Includes bibliographical references and index.
    ISBN 0-8493-2713-X
    1. Social sciences--Research--Methodology. 2. Humanities--Research--Methodology. 3. Geographic
information systems. 4. Spatial analysis (statistics) 5. Geographic information systems--Japan--
Databases--Case studies. I. Okabe, Atsuyuki, 1945-

H62.S7962 2005
300'.72'7--dc22   2005048572

**Visit the Taylor & Francis Web site at
http://www.taylorandfrancis.com**

**and the CRC Press Web site at
http://www.crcpress.com**

Preface

Almost all phenomena studied in the humanities and social sciences occur in geographical space. This implies that, in principle, studies in the humanities and social sciences can be enhanced by the use of geographical information systems (GIS). However, actually employing GIS in the advancement of these disciplines is not straightforward. Any computer-aided method of analysis is pointless unless researchers can devote the time necessary to learning what it is, what it can do, and how to use it. To this end, we carried out the six-year project entitled Spatial Information Science for the Humanities and Social Sciences (SIS for HSS). The project began in June 1998, when the Center for Spatial Information Science (CSIS) was established at the University of Tokyo, and ended in March 2004. The project was funded by the Grant-in-Aid for Special Field Research provided by the Ministry of Education, Culture, Sports, Science and Technology in Japan. The project leader was Atsuyuki Okabe of CSIS.

The SIS for HSS project had two aims:

1. To integrate spatial methods that were fragmentarily developed in the humanities and social sciences, in particular as applied to the areas of economics, human geography, and archaeology, and to develop the methods into GIS-based tools for studies.
2. To develop spatial data infrastructural systems that would support research in the above fields.

To achieve both of these objectives, the SIS for HSS project team had five groups, which are listed below with the name of each team leader. The first three of the groups were organized by subjects, and the last two were based upon the GIS technologies employed. All the groups worked in collaboration.

1. Economics (Yoshitsugu Kanemoto)
2. Human geography (Hiroyuki Kohsaka)
3. Archaeology (Takura Izumi)
4. Spatial data acquisition (Ryosuke Shibasaki)
5. Spatial data management (Yukio Sadahiro)

The achievements of the first objective, which are outlined in Chapter 1, are presented in 19 sections (Chapters 2–20 of this volume).

The achievements of the second aim were the development of:

- A spatial database that contains ready-to-use data commonly used in the humanities and social sciences
- A spatial-data clearinghouse in which researchers can easily search through spatial data in the database developed above at http://chouse.csis.u-tokyo.ac.jp/gcat/editQuery.do
- A data-sharing system that is widely used by scholars in the humanities and social sciences, www.csis.u-tokyo.ac.jp/japanese/research_activities/joint-research.html

These systems are run by CSIS, and are open to academic users. The systems are particularly useful when the researcher's interest is in studying human and social phenomena as they occur in Japan.

We sincerely hope that by means of this book, readers can come to an understanding of how GIS are actually utilized in advancing studies in the humanities and social sciences; furthermore, this book will encourage readers to develop new GIS-based methods in their own research.

Atsuyuki Okabe

Editor

Atsuyuki Okabe received his Ph.D. from the University of Pennsylvania in 1975 and his doctoral degree in Engineering from the University of Tokyo in 1977. Previously he has held the position of Associate Professor at the Institute of Socio-Economic Planning, University of Tsukuba. He is currently Professor of the Department of Urban Engineering, University of Tokyo, and served as Director of the Center for Spatial Information Science (1998–2005). His research interests include geographical information science, spatial analysis, spatial optimization and environmental psychology. He has published many papers in journals, books, and conference proceedings on these topics. He is a co-author (with Barry Boots, Kokichi Sugihara, and Sung Nok Chiu) of *Spatial Tessellations: Concepts and Applications of Voronoi Diagrams* (John Wiley). He edited *Islamic Area Studies with Geographical Information Systems* (RoutledgeCurzon). He serves on the editorial boards of many international journals, like the *International Journal of Geographical Information Science.*

Acknowledgments

So many people helped in very many ways during the preparation of this book that we are able to acknowledge only a few of them individually. First, we are deeply grateful to the Ministry of Education, Culture, Sports, Science, and Technology for financially supporting our project for six years. By coincidence, a similar, nationally funded project was undertaken in the United States by the Center for Spatially Integrated Social Science (CSISS) during virtually the same period. Exchange between the members of CSISS and those of SIS for HSS was fruitful. In particular, we express our thanks to Luc Anselin, Serge Rey, Nick Ryan, Stephen Matthews, and Gilles Duranton for commenting upon our studies in an international workshop. We also thank Tadaaki Kaneko for ably managing finances, documentation, Web pages, and symposia for six years. We are pleased to acknowledge the support of CSIS at the University of Tokyo, where the spatial-information infrastructure of our outcome is placed. Our special thanks go particularly to Tsuyoshi Sagara, Eiji Ikoma, Kaori Ito, Akiko Takahashi, Akio Yamashita, You Shiraishi, and Hideto Satoh. We are indebted to the staff of the publisher, especially Rachael Panthier, Jessica Vakili, Taisuke Soda, Tony Moore, Matthew Gibbons, and Randi Cohen. Finally, we also express our gratitude to Yoko Hamaguchi and Ayako Teranishi for preparing our manuscripts.

Contributors

Yoshio Arai
Department of Human Geography
School of Arts and Sciences
University of Tokyo

Masatoshi Arikawa
Center for Spatial Information
    Science
University of Tokyo

Yasushi Asami
Center for Spatial Information
    Science
University of Tokyo

Ali El-Shazly
Faculty of Engineering
Cairo University

Hidetomo Fujiwara
Graduate School of Frontier
    Sciences
Institute of Industrial Science
University of Tokyo

Naoko Fukami
Institute of Oriental Culture
University of Tokyo

Takashi Fuse
Department of Civil Engineering
University of Tokyo

Xiaolu Gao
Instutute of Geographyical
    Sciences and Natural Resources
    Research
Chinese Academy of Science

Yutaka Goto
Faculty of Humanities
Hiroaki University

Masashi Haneda
Institute of Oriental Culture
University of Tokyo

Yoshio Igarashi
Spatial IT Business Unit
Aerospace Division
Mitsubishi Corporation

Fumiko Itoh
Faculty of Economics
Niigata University

Yosinori Iwamoto
Graduate School of Frontier
    Sciences
University of Tokyo

Erina Iwasaki
Graduate School of Economics
Hitotsubashi University
Tokyo

Takura Izumi
Graduate School of Faculty of
    Letters
University of Kyoto

Yoshitsugu Kanemoto
Graduate School of Public Policy
    and Graduate School of
    Economics
University of Tokyo

Hiroshi Kato
Graduate School of Economics
Hitotsubashi University
Tokyo

Toru Kitagawa
Department of Economics
Brown University

Hiroyuki Kohsaka
Department of Geography
Nihon University

Shiro Koike
Department of Population Structure
    Research
National Institute of Population and
    Social Security Research

Yuki Konagaya
The National Museum of Ethnology
    Osaka
Japan

Reiji Kurima
Graduate School of Economics
University of Tokyo

Takanori Kimura
Services Delivery-Industrial
IBM Japan, Ltd.

Dinesh Manandhar
Center for Spatial Information
    Science, University of Tokyo

Atsushi Masuyama
Department of Real Estate Science
Meikai University

Susumu Morimoto
Nara National Cultural Properties
    Research Institute

Yoshiyuki Murao
GIS Business Promotion
IBM Japan

Masafumi Nakagawa
National Institute of Advanced
    Industrial Science and
    Technology

Katsuyuki Nakamura
Center for Spatial Information
    Science
University of Tokyo

Izumi Niiro
Department of Archaelogy
Okayama University

Atsuyuki Okabe
Center for Spatial Information
    Science
University of Tokyo

Kei-ichi Okunuki
Department of Geography
Graduate School of Environmental
    Studies
Nagoya University

Saiko Sadahiro
Faculty of Education
Chiba University

Yukio Sadahiro
Department of Urban Engineering
University of Tokyo

Hiroshi Saito
Department of Economics
Tokyo University

Tomoko Sekine
Department of Geography
Nihon University

Ryosuke Shibasaki
Center for Spatial Information
  Science
University of Tokyo

Eihan Shimizu
Department of Civil Engineering
University of Tokyo

Keiji Shimizu
GIS Division
Kanko Co., LTD

Shino Shiode
Center for Spatial Information
  Science
University of Tokyo

Etsuro Shioji
International Graduate School of
  Social Sciences
Yokohama National University

Hiroya Tanaka
Faculty of Environmental
  Information
Keio University

Takashi Tominaga
Industry Business Unit
Region Metro
Small and Medium Business
IBM Japan, Ltd

Hiro'omi Tsumura
Faculty of Culture and Information
  Science
Doshisha University

Teruko Usui
Department of Geography
Nara University

Tohru Yoshikawa
Faculty of Urban Environmental
  Sciences
Tokyo Metropolitan University

Huijing Zhao
Center for Spatial Information
  Science
University of Tokyo

Table of Contents

1

Introduction

Atsuyuki Okabe

CONTENTS

1.1   What Are Geographical Information Systems (GIS)?

We notice in the literature of the humanities and social sciences that many studies deal with phenomena that are closely related to geographical factors. For example:

- Population change over 100 years is related to change in the net-works of arterial roads and railways (Chapter 5).
- Travel behavior in a 17th century city was related to the configuration of landmark buildings (Chapter 11).
- Configuration of ancient tax regions was related to fishing and agricultural areas (Chapter 12).
- Size of paleo-settlements was related to hunting and fishing localities (Chapter 13).
- Migration behavior is related to low-income regions (Chapter 14).
- Housing prices are related to the surrounding environment (Chapter 15).
- Agglomeration economies are related to city size (Chapter 16).

- School systems are related to the areal configuration of elementary and lower secondary schools (Chapter 17).
- Clinic service areas are related to the travel time of the patients (Chapter 18).

Groupings of these phenomena that are closely related to geographical factors are called *geographical phenomena*.

Traditionally, researchers in the humanities and social sciences study geographical phenomena with the aid of paper maps, and most of their tasks are undertaken by hand. For instance, they count the number of archaeological sites in a region by marking each site on a map with a pencil; they then measure the distance between sites by placing a ruler on a map; they then measure the area of each site by counting the number of grid cells covered by a transparent grid sheet placed over the map; then the slope angles of an archaeological site are determined by counting the number of contour lines; and so forth. Such tasks are tolerable when the number of geographical features is small, but once these variables become numerous, the work is laborious and time consuming. This difficulty is one of the reasons why geographical factors, despite their significance, have often been ignored in the study of humanities and social science.

Fortunately, in the late 1980s, user-friendly, computer-based processing tools, called *geographical information systems*, became available, and these greatly assisted in overcoming the tedious and time-consuming tasks. GIS are, in short, computer-based methodologies for processing geographical data.

What follows describes the key terms. *Geographical data* refers to the data on geographical features and consists of *spatial-attribute data* — the locational and geometrical attributes of features — and *nonspatial-attribute data* — attributes other than spatial ones. Geographical data are alternatively called *spatial data*. The difference is subtle, but geographical data usually refer to the ground surface (two-dimensional), while spatial data may include information on the ground surface and also three-dimensional observations for above and below ground, such as atmospheric and ground-water conditions. Furthermore, geographical recordings may not include measurements of architectural space, while spatial data include these. Since this book includes the data relevant to archaeological buildings, railway-station halls, and similar cultural and social constructions, the term *spatial data* is preferred, and mainly used.

The second key term in our consideration of GIS is *processing*. This refers to the application of the following subprocesses to the spatial data:

1. Acquiring
2. Managing
3. Analyzing
4. Visualizing

A full explanation of these procedures would require a dedicated book, but here, in Section 1.1, we briefly explain subprocessing for the convenience of readers who are not familiar with GIS. Others more familiar with GIS may ignore this part and go to Section 2. Please note that a 17-page introduction to GIS is provided by Okabe (2003).

The first step in subprocessing, i.e., acquiring spatial data, is classified into "direct" and "indirect" acquisition.

Direct spatial-data acquisition means observing and recording entities in the real world, for example, taking pictures of houses with established geographical locations and dimensions (Chapter 2); scanning of archaeological evidence by laser scanner (Chapter 3); tracing the trajectories of moving people in a station hall by laser scanner (Chapter 4); imaging land cover by airborne remote-sensing equipment mounted on airplanes and satellites; interviewing immigrants to determine their origins in field surveys (Chapter 14); and so forth.

Indirect spatial-data acquisition means deriving spatial data from material represented by conventional maps and census documents that contain information obtained from direct observations, such as administrative boundaries defined by surveying and set down as part of a map. In this process, electronic scanning employing a device like a facsimile or tracing the boundaries of features by a digitizer (a computerized device for tracing) may be done. Imputation of population data for villages recorded in a census book and the association of rural boundaries and their populations (Chapter 5) may be undertaken by computer, and so forth.

The second step in subprocessing, i.e., managing spatial data, is organizing the acquired data so they can be easily retrieved and manipulated. A system for this subprocessing is called the *spatial database*. This consists of two components: first, a database for spatial attributes, which manages geometrical and locational data of features, and second, data on nonspatial characteristics. Methods differ according to the data types, which are "raster" and "vector."

Raster data represent features in terms of *pixels*, which are dots or squares arrayed on a rectangular lattice with attribute values placed on each pixel. A good example is remotely sensed data (Figure 1.1), which appears in picture form at a distance (Figure 1.1a) while the squares constituting the images become visible on zooming in (Figure 1.1b). Data are simply managed through an array of numbers representing attribute values and the coordinates of pixels (Figure 1.1c).

Vector data represent features in terms of points, line segments, and polygons (Figure 1.2a). These geometrical elements are recorded as the coordinates of points, the names of start and end points for line segments, and, counterclockwise, the names of vertices for polygons. Management of vector data is not as simple as for raster data when we wish to know the topological properties within points, line segments, and polygons. That is, which line segments cross another given line segment, which polygon includes a certain point, and which polygons are adjacent to a particular polygon?

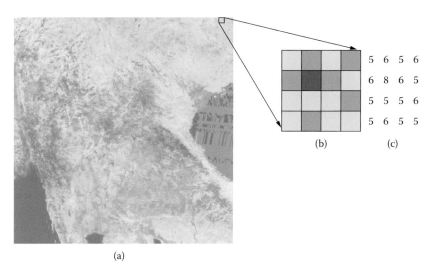

FIGURE 1.1
Raster data: (a) zoomed out, (b) zoomed in, and (c) their array of values.

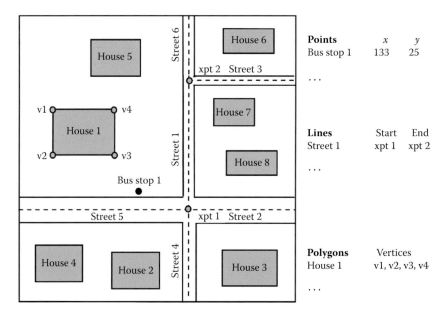

FIGURE 1.2
Vector data: (a) geometrical elements (points, lines, and polygons), and (b) the related numerical data for points, lines, and polygons.

The underlying theories for managing topology are fairly complicated, but users can easily use an ordinary GIS without knowing the underlying theories. The database for nonspatial attributes usually adopts a table-type format called a *relational database*. Frequently used examples of this are

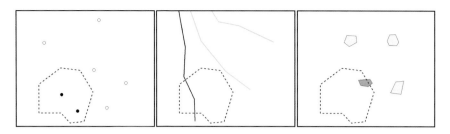

FIGURE 1.3
Inclusion search operations for (a) points, (b) line segments, and (c) polygons that are partly included in a given polygon (indicated by the broken line).

Microsoft Excel and Microsoft Access. Readers wishing to understand the supporting theories of spatial databases should consult, for example, Shekhar and Chawla (2003).

The third step in subprocessing is the analysis of spatial data. This is the main function of GIS, providing many operations for analysis of spatial, as well as nonspatial, data. Since the steps needed for the analysis of nonspatial-attribute data are fairly well-known, such as the operations in Excel, we will focus on the analysis of spatial-attribute data. A first set of operations is engaged to measure geometrical quantities. Examples are the measurement of the distance between two points, of the length of a line consisting of straight segments, of the area of a polygon, of the angle of a slope, and so forth.

A second set of operations is used for spatial searches. Frequently used approaches are the "inclusion search," "distance search," and "intersection search." The *inclusion search* finds those points, lines, and polygons that are partly included in a given polygon. For example, these searches are used for finding hospitals (the points in Figure 1.3a), streams (the line segments in Figure 1.3b), and parks (the continuous-line polygons in Figure 1.3c) in a given area (the broken-line polygon in Figures 1.3a, b, and c).

The *distance-search* operation (which is closely related to the "buffer" process to be shown later) finds points, line segments, or polygons, parts of which are within a given distance from a given geometrical element (Figure 1.4). For example, these searches are used to locate hospitals (Figure 1.4a), streams (Figure 1.4b), and parks (Figure 1.4c) that are within 200 meters from an expressway (the dot–dash lines in Figures 1.4a, b, and c).

The *intersection-search* operation finds line segments or polygons that intersect with given similar elements (Figure 1.5). For example, it finds streams that intersect with an expressway (Figure 1.5a) or, similarly, parks (Figure 1.5b).

A third manipulation is called the *buffer* operation, which generates a new area in which the distance to the nearest feature is within a given distance from given geometrical elements. For example, the buffer operation for point-like features, such as stations, gives the area in which the distance to the nearest station is within a certain limit, say, 200 meters (Figure 1.6a). The buffer operation for line-like features, such as streams, reveals the area in

FIGURE 1.4
Distance search operations for (a) points, (b) line segments, and (c) polygons, part of which are within 200 meters from a given line indicated (the dot–dash lines).

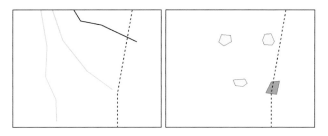

FIGURE 1.5
Intersection search operations for (a) line segments and (b) polygons that intersect with the given broken line.

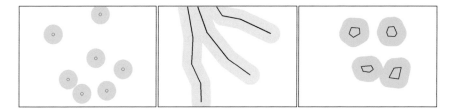

FIGURE 1.6
Buffer operations for (a) points, (b) line segments, and (c) polygons.

which the distance to the nearest point on the streams is within a certain distance (Figure 1.6b). The same process applied to an area-like feature, such as a park, generates the area in which the distance to the nearest point on the park's boundary is within a certain distance (Figure 1.6c).

A fourth set of operations, called the *overlay* operation, generates a new spatial-data set by overlaying two different spatial data sets. Many processes are included in the overlay operation. Three of the most frequently used are OR (union), AND (intersection), and NOT (compliment). To take examples, suppose that A1 contains the areas in which the distance to the nearest hospital is within 200 meters (Figure 1.6a), and A2 contains the areas in which the distance to the nearest point on streams is within 200 meters

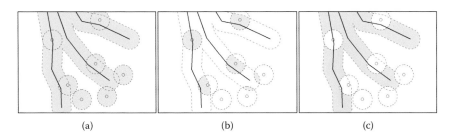

FIGURE 1.7
Overlay operations: (a) OR, (b) AND, and (c) NOT.

(Figure 1.6b). Then, the operation A1 OR A2 generates a new area in which the distance to the nearest hospital is within 200 meters, or the distance to the nearest point on a stream, as shown in Figure 1.7a. The operation A1 AND A2 is shown in Figure 1.7b, and A2 NOT A1 is shown in Figure 1.7c.

Combining these basic operations of GIS, we can analyze spatial data. In addition, GIS provide tools for advanced methods called *spatial analysis*, which include *spatial statistics*. Tools for spatial analysis and statistics are shown in Part 3, and their applications are shown in Part 4 of this volume. Good textbooks are Bailey and Gatrell (1995) for spatial analysis and Cressie (1993) for spatial statistics.

The last category of subprocessing is visualizing spatial data, which is the outcome of spatial analysis. Ordinary GIS provide many tools for visualization. To make an attractive and easily understandable visual product, usually in the form of maps, we have to consider several characteristics of spatial data: the geometrical types of features (e.g., points, lines, polygons, solids, etc.), the measurement scales of attribute values (e.g., nominal, ordinal, interval, ratio scales, etc.), spatial-data units (e.g., feature-based units, tessellations, cell grids, continuous space, etc.), and other features. Considering these characteristics, we develop a visual product of what we wish to convey in terms of visual variables (e.g., spacing, size, shape, hue, lightness, arrangement, etc.). For details, see, for example, Slocum (1999).

Visualization achieved through the use of GIS tools has much variety, and so we can freely enjoy this. But sometimes we want to visualize spatial data in a conventional fashion. Four of the most conventional map methods are "choropleth," "proportional symbol," "isarithmic," and "dot."

A *choropleth map* is made by shading the cells of a tessellation, with an intensity proportional to attribute values. An example is shown in Figure 1.8a, which illustrates the number of street robberies that occurred in districts of Saitama Japan.

A *proportional symbol map* is made by scaling symbols in proportion to the magnitude of an attribute value of a feature located at a representative point. An example is shown in Figure 1.8b, which is an alternative presentation of Figure 1.8a.

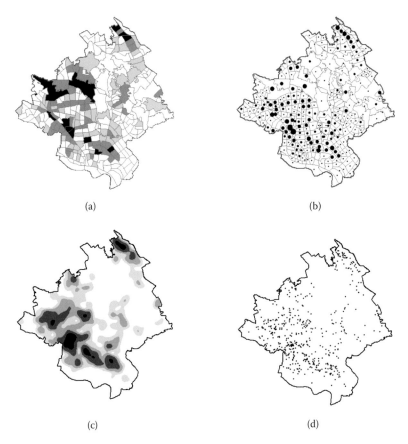

(a)   (b)

(c)   (d)

FIGURE 1.8
Street robberies in Saitama represented by different map types: (a) choropleth, (b) proportional symbol, (c) isarithmic, and (d) dot map.

An *isarithmic map*, which is alternatively called a *contour map*, is based upon a set of lines, called *isolines*, joining the same attribute values. An isarithmic map is usually obtained from the density function estimated from known values at finite points. Note that this procedure is called *spatial interpolation*. An example is shown in Figure 1.8c, which illustrates the locational density of street robberies in Saitama.

A *dot map* is a set of points located on a plane, with each point representing the place of an event, for example, the site of a crime. An example is shown in Figure 1.8d, which shows the locations of Saitama street robberies.

The above is an outline of the components of GIS. Readers who wish to know GIS methods in more detail should consult textbooks, for example, Bernhardsen (2002), Burrough and McDonnell (1998), Clarke (2003), Christman (2002), Delaney (1999), Demers (2000), Heywood et al. (2002), Jones (1996), Lo and Yeung (2002), Longley et al. (2001), Wise (2002), and Worboys (1995).

1.2 Applications of GIS in the Humanities and Social Sciences: Overview of the Chapters

Having understood what GIS are in Section 1.1, readers must now realize that GIS are potentially very useful. As a matter of fact, this volume shows how GIS are valuably applied to various studies in the humanities and social sciences. The volume consists of 20 chapters, including this introductory chapter. The subsequent 19 chapters are classified into five parts:

Part 1. Spatial-data acquisition

Part 2. Spatial databases

Part 3. Tools for spatial analysis

Part 4. Applications of spatial analysis

Part 5. Visualization

These sections cover almost all the basic components of GIS, which correspond to the four subprocesses within GIS mentioned in Section 1.1. Explicitly, Part 1 deals with the acquisition of spatial data; Part 2 considers data management; Parts 3 and 4 examine analysis; and Part 5 looks at visualization. Through reading this volume, readers can therefore understand how GIS are actually applied to studies in the humanities and social sciences.

Part 1, Spatial-Data Acquisition, consists of Chapters 2, 3, and 4. Chapter 2 introduces one of the simplest methods of acquiring this information, namely, taking photographs, which are a useful medium for establishing a record of places, people, life, and the atmosphere. It is not unusual for observers to take more than 100 pictures per day in a field study. However, when a researcher comes back from a field survey, he/she is often at a loss when it comes to organizing a heap of images on the desk.

It is particularly hard to reproduce a three-dimensional space using photographs. To overcome these difficulties, Chapter 2 discusses a good tool called *STAMP*. This method has been developed from two techniques, "photo collage" and "hypermedia." *Photo collage* is a picture made by a combination of bits of photographs. *Hypermedia* is a system for linking two pages by a hyperlink, which, readers will recall, is commonly used in linking Web pages. Combining these two techniques, STAMP integrates many photographs in a quasi-three-dimensional geographical space through which we can virtually walk and thus experience. Chapter 2 also demonstrates an application of STAMP to an ethnographic study in Mongolia. Note that STAMP is downloadable without charge via the Internet.

Chapter 3 introduces one of the high-tech methods for acquiring three-dimensional data, namely, laser scanners. Having heard the term "laser scanner," one might recall a pointer of light used for highlighting a specific place on a PowerPoint slide. In principle, the laser scanner discussed in

Chapter 3 (and also in Chapter 4) is similar to this, although the former is a more advanced device that measures the round-trip time between a laser and a shot point (spot) on the surface of an object. The three-dimensional coordinates of the spot are estimated from the travel time, the angle of the beam from the laser, and its location. By sweeping the beam over the surface, the laser scanner obtains the data from the spots, called the "point-cloud" data of the object, which provide the three-dimensional digital data after editing. Chapter 3 illustrates this data-acquisition method in an easy-to-understand manner. This chapter also describes a system for collecting and organizing archaeological data, called *Archae-Collector*, it greatly helps scholars acquire, organize, and share data among an excavation team, even during the excavation work.

Chapter 4 also shows a method for acquiring spatial data with laser scanners. A distinct difference between the methods in Chapter 3 and Chapter 4 is that the former acquire the spatial data of stationary objects, whereas the latter determine those of moving forms. Chapter 4 considers a laser-based system for recording the trajectories of pedestrians. This method is easier and more precise than the conventional approach using a video camera. A key technique of the new method is the ability to identify the trajectory of the same pedestrian. Imagine many pedestrians walking in many directions in a railway-station hall during the rush hour. This technique is realized through a pedestrian-walking model using the Kalman filter. The developed system was installed on a railway-station concourse, and almost 100 percent accuracy was achieved for a spatial density of less than 0.4 persons per square meter. There are many potential applications in behavioral science, sociology, environmental psychology, and human engineering.

Part 2, Spatial Databases, consists of Chapters 5, 6, and 7. Chapter 5 deals with a historical population database. To study a structural change in the population distribution of a region, population data covering 100 years are necessary. However, such long time-span population records are usually not available. Chapter 5 shows how to reconstruct historical population data from ancillary sources. One of the most useful of these supportive information sources is old maps. These old manuscripts do not show population, but they illustrate the distribution of houses. A problem is how to convert the areas occupied by houses into the number of inhabitants. Chapter 5 finds an empirical function for this conversion based on the correspondence between the areas occupied by houses in a district and an old document showing the population in the same area. Using this function, Chapter 5 reconstructs the population grid data of the Kanto Plain for 1890 and 1930. These data sets are integrated in the existing population-grid data sets of 1970 and 2000, and a 110-year population-grid database is constructed. Using this database, Chapter 5 shows the structural change of population distribution in the Kanto Plain over a period of 100 years. This population database is accessible via the Internet.

Chapter 6 deals with a statistical database for urban areas. In urban economics, such data are indispensable, but a problem exists in that there is no precise definition for urban areas. The legal definition of a city is often used for an urban area, but many activities extend beyond jurisdictional boundaries, and legal "urban areas" are different from the actual ones. The federal government of the U.S.A. has been trying to define actual urban areas since 1947. These are designated Standard Metropolitan Statistical Areas, Consolidated Metropolitan Areas, and Core-Based Statistical Areas. First, the central cities are defined, and second, the suburban areas for each are formally identified. However, this way of definition has become increasingly problematic in recent years, because a large number of subcenters have been recognized to have emerged, and commuting patterns have become increasingly complex.

Chapter 6 proposes a new iterative method for defining urban areas using GIS called *urban employment areas*. The chapter considers a spatial database constructed by applying this method, which includes the numbers of employees and populations in 1980, 1985, 1990, and 1995; production (value added); and private-capital stocks and social-overhead capital. This database is accessible via the Internet.

Chapter 7 discusses the methods used in constructing a universal database for archaeological observations. Generally speaking, one of the most difficult problems encountered with GIS is spatial-data transfer among different researchers, communities, and GIS software. Archaeological data are no exception. To overcome this difficulty, a technical committee of the International Standard Organization, namely ISO/TC211, has proposed a data-transfer standard that is being increasingly accepted by many countries and that has actually become an international standard. However, this standard is too general to manage particular research disciplines, as exemplified by the need to accommodate the complexity of archaeological artifact characteristics.

Based on a critical examination of traditional European as well as Japanese methods, Chapter 7 proposes an object-oriented spatial database for managing archaeological data in terms of the Unified Modeling Language (UML). The object-oriented spatial database is distinct from the *layer-based* one that manages spatial data with a collection of map sheets, for instance land use, road, and railway maps, among others. The *object-oriented* spatial database holds spatial data as an assemblage of geographical features that are characterized by their classes and relationships. The *UML* is a widely used language for describing object-oriented spatial databases in terms of pictorial elements, such as squares, lines, arrows, and other features. Chapter 7 demonstrates how to construct a spatial database for the management of feature descriptions in archaeological sites using UML.

Part 3, Tools for Spatial Analysis, consists of Chapters 8, 9, and 10. Chapter 8 shows how to locate tools for spatial analysis via the Internet. As mentioned in Section 1.1, the ordinary GIS software provides many basic tools for spatial analysis, but they are not always sufficient to analyze specific situations in the humanities and social sciences. Fortunately, a considerable number of tools for advanced analysis have been developed by the GIS community, and information about these is posted on the Internet. However, such information is scattered across the Web, and it is difficult to find an appropriate tool for a specific application. In fact, Google shows more than 3 million Web sites referring to spatial analysis. Chapter 8 introduces two Web sites that provide appropriate search engines. The first is served by the Center for Spatially Integrated Social Sciences. The second, called *FreeSAT,* provides for the locating of free software packages for spatial analysis. Both sites are easily accessible via the Internet.

Chapter 9 illustrates how to use a toolbox for examining the spatial effect of features on the distribution of events. In the real world, we notice many events that occur at specific locations. These are called *spatial events*, and they include the location of facilities in particular places. Spatial events are in part affected by their constraining geography, in particular by influencing elements that persist over a long time period. These durable controls are called *infrastructural features*. Examples of these that have attracted research in the humanities and social sciences are:

- Transport stations attracting crime in Los Angeles
- Mosques being usually located on hilltops in Istanbul
- Asthma sufferers residing 200–500 meters from major highways in Erie County, New York
- Baltimore serial thieves having a tendency to migrate south along the major roads

Chapter 9 introduces a user-friendly toolbox, called *SAINF,* which may be used in the statistical analysis of these spatial relationships. SAINF can be downloaded via the Internet without charge.

Chapter 10 demonstrates how to use a toolbox called *SANET* for analyzing spatial events that occur in a network or alongside a network. These are referred to as *network spatial events*. Some typical examples relevant to studies in the humanities and social sciences are:

- Homeless people living on the streets
- Street crimes
- Graffiti sites along highways
- Traffic accidents
- Street-food stalls
- Fast-food stores in a downtown

The toolbox introduced in Chapter 10 is useful for answering, for instance, the following questions:

- Does illegal parking tend to occur uniformly in no-parking streets?

- Are street-crime locations clustered in "hot spots"?
- Do fast-food shops tend to compete with each other?
- What is the probability of consumers choosing a particular down-town store?

Chapter 10 illustrates how to use the tools of SANET for finding answers to these questions. SANET can be downloaded via the Internet without charge.

Part 4, Applications of Spatial Analysis, consists of seven chapters, which show spatial analyses in history (Chapter 11), archaeology (Chapters 12 and 13), sociology (Chapter 14), housing economics (Chapter 15), urban economics (Chapter 16), and educational administration (Chapter 17).

Chapter 11 presents an application of spatial analysis, or, in specific terms, "spatial reasoning," to a study in history. Historical facilities often reveal historical evidence, and their locations are of particular interest. If there are maps exactly indicating the locations of facilities, there will be no need for a locations search. However, such historical maps are often unavailable, and, even if they exist, a number of facilities may not be recorded on the maps. In such cases, historical documents, if any, are only a means of inferring the location. For this purpose, spatial reasoning can be of potential use. *Spatial reasoning* is an attempt to infer the unknown locations of features and their relationships from appropriate known sites. Chapter 11 illustrates the applicability of spatial reasoning in historical analysis by its application to the inference of spatial locations written about in Jean Chardin's travel account and his walking route in Isfahan in Iran in the 17th century.

Chapter 12 shows spatial analysis used in archaeology. In the eighth century, much of Japan was ruled within a hierarchy of administrative districts called the *go-ri* system. A *go* comprised 50 houses (called *ko*), and this was divided into two or three *ris*. On average, a *go* contained more than 1000 persons. There has been a long debate over whether the *go* and *ri* show the actual villages and families at the time or whether they were predominantly contrived by the authorities. Most scholars agree that administrative influence was strong, but opinions differ over the extent to which the divisions reflect the reality of ancient Japan. Chapter 12 attempts to answer this question by reconstructing agricultural productivity in the West Wakasa region using GIS.

Chapter 13 also shows spatial analysis, or, in specific terms, "site-catchment analysis," in archaeology. In the late 20th century, the Sannai-Maruyama site (5900 to 4200 BP) was excavated in the northern part of Japan. This site is distinctly different from others in Japan in two respects. First, the number of dwellings was greater than the archaeologists first considered. Fifty to 100 houses were discovered in one archaeological phase, suggesting that 200 to 400 hundred people lived together. Second, the life span of villages was much longer than first believed. Most villages were maintained for one to three generations (50–100 years), and people lived continuously at this site for 1700 years. The questions arose: What kind of subsistence strategies supported such large-sized and long-lasting villages? What kind of environmental elements allowed the inhabitants to live such a lifestyle? Chapter 13 answers these questions by reconstructing the paleo-environment using a GIS.

Chapter 14 shows spatial analysis, otherwise considered as migration analysis, in sociology. A few decades ago, most sociological studies on Egypt showed that Egyptian rural society was fairly homogeneous, in the sense that people tended to move from all the rural areas to the urban areas, and this trend did not differ between rural districts. However, this observation was based on macro data that were the only available information source at that time because of military and security restrictions. After the establishment of the Egyptian open-economic policy, these restrictions were removed, and it became possible to conduct surveys to acquire micro social and geographical data. This chapter analyzes the migration behavior from rural to urban areas by integrating census data, detailed household survey information, and geographical observations through a GIS, and reveals a significantly different outlook to the former one.

Chapter 15 examines spatial analysis as a hedonic price analysis in housing economics. In big cities, an increase in detached-housing density causes various environmental problems, such as a decrease in vegetation, obstruction of sunshine, insufficient ventilation, and other effects. To help resolve such environmental problems, it is necessary to study the effects of microlevel residential environments on the housing price. A decade ago, this study was difficult to undertake because there were few microlevel data, and it was difficult to manipulate detailed geographical data together with the topology and achieve a result from the interface of features. GIS has resolved this difficulty, and using GIS and computer aided design (CAD), Chapter 15 estimates a housing-price function of microlevel residential environmental factors. The effects of policies that affect the subdivision of lots, the construction of pocket parks, road widening, relaxation of the regulated ratio of the area of a building floor to the area of a lot, and other factors are considered.

Chapter 16 looks at spatial analysis in urban economics. Tokyo, with a population exceeding 30 million people, is the largest city in Japan. There has long been debate about whether Tokyo is too large. Undesirable aspects of Tokyo might be densely packed commuters in rush-hour trains, heavy traffic congestion on expressways, car exhaust emissions, and many other factors of overcrowding. At the same time, Tokyo provides a good business environment where face-to-face communication is easily realized. A problem is how to measure the economies or the diseconomies of Tokyo. Chapter 16 considers the size of agglomeration economies using the GIS database of the Metropolitan Employment Area constructed in Chapter 6, and tests whether Tokyo is too large by applying the Henry George Theorem.

As seen in Chapters 15 and 16, spatial analysis includes spatial optimization or, to some extent, the evaluation of spatial policies. Chapter 17 deals with the evaluation of school redistricting based on a study of the "school family system," which has been advocated in recent years. The *school family system* means that one elementary school cooperates with several lower secondary schools in educating students. Introduction of this system to the existing conventional system, in which elementary and lower secondary schools are independent, results in a school-redistricting problem. Chapter 17 considers this problem by resolving the optimization problem of minimizing the average distance from home to school and the number of students assigned to different schools. Chapter 17 also shows an application of this method to an actual school system in Kita, Tokyo.

Part 5, Visualization, consists of Chapters 18, 19, and 20. Chapter 18 shows a three-dimensional visualization method for historical studies. The landscapes of old cities reproduced from old maps, pictures, and documents provide a new tool for understanding the living environment of ancient people. In the reproduced three-dimensional city, researchers are able to live virtually among the people of that time, and they can observe the prevailing environmental ambience of that time by going along the city streets using a walk-through simulation. However, it is difficult to reproduce a three-dimensional city from old materials, because these are usually imprecise with respect to location. A problem is how to impose vague spatial information on precise, three-dimensional information. Chapter 18 gives a way of solving this problem and applies it to old Tokyo, called Edo.

Chapter 19 deals with the visualization for "site assessment" often discussed in marketing. *Site assessment* is the evaluation of factors affecting the choice of appropriate sites for facilities. One of the most important considerations is the accessibility of sites. This is usually measured using a certain size of spatial-data units, such as districts or grid cells. However, the value given to accessibility dramatically changes according to the choice of such units. This notorious problem is called the *Modifiable Area Unit Problem* (MAUP) in spatial analysis. If large units (coarse units) are used, the distribution of accessibility values shown in a map is easily grasped, but inaccuracy increases. If small units (fine units) are used, accuracy is gained but detailed accessibility values are difficult to grasp visually. Chapter 18 shows how to visualize accessibility values and their levels of accuracy at the same time.

The last chapter, Chapter 20, deals with the visualization of the personal impression or the perceived image of geographical space, which is sometimes called a *mental map*, as in environmental psychology. The mental image of a geographical feature, say, a district and a street, is usually communicated verbally using adjectives. For example, "That street is lively and elegant." Visualization of this impression is difficult, because the measurement of the subjective magnitude of "liveliness" or "elegance" is harder than that of an objective value, such as rainfall. Moreover, the subjective magnitude varies from person to person, from place to place, and from daytime to nighttime. Chapter 20 develops a method for visualizing the subjective impressions of districts in a city with a three-dimensional GIS and applies it to the visualization of "liveliness" and "elegance" over space and time in Shibuya, Tokyo.

As may be noticed from the above overview of the following chapters, GIS are invaluable for modern studies in the humanities and social sciences. Needless to say, applications of GIS are not limited to the above subjects, and the potential applications are numerous. Actually, in recent years, books dealing with GIS-based studies in the humanities and social sciences have been published, for example, Okabe (2004) in Islamic-area studies, Wheatley and Gillings (2002) in archaeology, Kindner et al. (2002) in socioeconomics, Leipnik and Alvert (2003) in law, Chainey and Ratcliffe (2005) in criminology, Craig et al. (2002) in public participation, and Wheatley and Gillings (2004) in health science. We expect that readers who study this volume will develop new applications in their own fields.

Acknowledgments

We express our thanks to W. Takeuchi for making Figure 1.1, T. Sato for Figures 1.3, 1.4, 1.5, 1.6, and 1.7, and M. Ichimura for Figure 1.8.

References

Bernhardsen, T., *Geographic Information Systems: An Introduction*, John Wiley & Sons, New York, 2002.

Bailey, T.C. and Gatrell, A.C., *Interactive Spatial Analysis*, Longman, Harlow, 1995.

Burrough, P. A. and McDonnell, R. A., *Principles of Geographical Information Systems*, Oxford University Press, Oxford, 1998.

Chainey, S. and Ratcliffe, J., *GIS and Crime Mapping*, John Wiley, London, 2005.

Chrismann, N., *Exploring Geographic Information Systems*, 2nd ed., John Wiley, New York, 2002.

Craig, W.J., Harris, T.M., and Weiner, D., *Community Participation and Geographic Information Systems*, Taylor & Francis, London, 2002.

Cressie, N., *Statistics for Spatial Data*, John Wiley, New York, 1993.

Clarke, K.C., *Getting Started with Geographic Information Systems*, 4th ed., Prentice Hall, Upper Saddle River, New Jersey, 2003.

Delaney, J., *Geographical Information Systems: An Introduction*, Oxford University Press, Melbourne, Australia, 1999.

Demers, M. N., *Fundamentals of Geographic Information Systems*, 2nd ed., John Wiley, New York, 2000.

Heywood, I., Cornelius, S., and Carver, S., *An Introduction to Geographical Information Systems*, 2nd ed., Prentice Hall, London, 2002.

Jones, C.B., *Geographical Information Systems and Computer Cartography*, Pearson Education Limited, Harlow, 1996.

Kidner, D., Higgs, G., and White, S., Eds., *Socio-Economic Applications of Geographical Information Science*, Taylor & Francis, London, 2002.

Leipnik, M. and Alvert, D., *GIS in Law Enforcement: Implementation Issues and Case Studies*, Taylor & Francis, London, 2003.

Lo, C.P. and Yeung, A.K.W., *Concepts and Technique of Geographic Information Systems*, Prentice Hall, Upper Saddle River, New Jersey, 2002.

Longley, P.A., Goodchild, M.F., Maquire, D.J., and Rhind, D.W., *Geographic Information Systems and Science*, 2nd ed., John Wiley, London, 2005.

Maheswaran, R. and Craglia, M., Eds., *GIS in Public Health Practice*, CRC, Boca Raton, Florida, 2004.

Okabe, A., Ed., *Islamic Area Studies with Geographical Information Systems*, Routledge Curuzon, London, 2004.

Shekhar, S. and Chawla, S., *Spatial Databases: A Tour*, Prentice Hall, Upper Saddle River, New Jersey, 2003.

Slocum, T.A., *The Thematic Cartography and Visualization*, Prentice Hall, Upper Saddle River, New Jersey, 1999.

Wheatley, D. and Gillings, M., *Spatial Technology and Archaeology: The Archaeological Applications of GIS*, Taylor & Francis, London, 2002.

Wise, S., *GIS Basics*, Taylor & Francis, London, 2002.

Worboys, M. F., *GIS: A Computing Perspective*, Taylor & Francis, London, 1995.

2

A Tool for Creating Pseudo-3D Spaces with Hyperphoto: An Application in Ethnographic Studies

Hiroya Tanaka, Masatoshi Arikawa, Ryosuke Shibasaki, and Yuki Konagaya

CONTENTS

2.1 Introduction

In the field of ethnography, researchers deal with a large number of photos. A researcher's mission is to stay in developing countries, often in Asia or Africa, and describe their environment while sharing experiences with native people. During their stay, they try to record their everyday life in detail (Konagaya, 1991; Konagaya, 1998; Konagaya, 1999). Some researchers will stay for about one year in the target locale and take more than 100 photos per day. Photos are a particularly useful medium for establishing a record of places, people, life, and atmosphere in a target area because they provide a rich source of information about spaces or places in the real world. The main advantages of photos in such studies are:

- Instant acquisition: An ordinary camera can be used quickly and easily.
- Rich information: Far from representing abstract data, photos are genuine records of real-world phenomena.
- Subjectivity: A photo often reveals not only the photographer's target, but also the subjective point of interest (POI) of the photographer.
- Scalability: Photos include micrographs, landscape photos, satellite photos, and so on. Photos thus enable us to examine the real world on any spatial scale.

While photos are clearly the most important medium for such studies, several problems make it difficult to effectively manage large collections of photos. These problems are related to two aspects of the way we manage photos:

- Organization and association
- Browsing and interaction

A few methods for organizing and associating photos have already been developed. For example, one method is to plot locations of photos onto two-dimensional (2D) maps like a typical geographical information systems (GIS) (Figure 2.1). If the location (latitude and longitude) where each photo is taken can be recorded using global-positioning system (GPS) and other technologies, we can plot locations of photos onto a map almost automatically. However, such conventional methods do not enable us to sufficiently understand the *relationships* among photos. With regard to browsing and interaction, two general approaches are usually taken — the album approach and the slide show approach (Figure 2.2). However, neither of these is essentially interactive or sufficient to describe the dynamic experiences of the photographer.

In this chapter, we describe a new method for managing photos through computers. We have developed and implemented a software that we call STAMP (Spatio-Temporal Association with Multiple Photos) [Tanaka (2002a), Tanaka (2002b), Tanaka (2003)]. STAMP allows us to easily create, publish, share, and navigate pseudo-three-dimensional (3D) spaces comprised of multiple photos. With this software, users can create not only intensive scenes, but also extensive 3D scenes, which enable interactive and continuous navigation. Furthermore, users can relive their own or other's experiences sequentially. Researchers can manage photos based on the *context* of the photos by using this software. Our software can solve

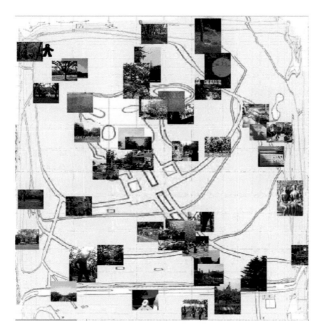

FIGURE 2.1
Photos plotted their locations onto a 2D map.

FIGURE 2.2
An album approach and a slide show approach.

several of the problems mentioned above and advance the use of photo applications.

In the following sections, we describe two important concepts that under-lie our approach — photo collage and hypermedia. In Section 2.3, we explain STAMP and discuss the characteristics of STAMP. In Section 2.4, we report on an experiment using STAMP, and then conclude in the final section.

FIGURE 2.3
An example of 2D photo collage.

2.2   Related Concepts

2.2.1   Photo Collage

Our approach is based on artistic representation through the concept of photo collage on a 2D canvas; that is, a general method of organizing and associating photos. People can connect, overlap, and rotate photos on the 2D canvas to compose a single image that represents a memorable sight or event in the real world. Figure 2.3 shows an example of our photo collage.

David Hockney, a famous artist who created so many photo collages, said that photo collage clearly reveals the photographer's perception of the world (Hockney, 1985). David named the series of his photo collage works "Moving Focus." This naming implies his notion about photo collage.

Photo collage is an effective way to assemble multiple photos into one complete image. Moreover, we can read through multiple scenarios in a single photo collage. Figure 2.4 shows an example of the visual paths in our photo collage.

The photo collage approach has shortcomings, though, in that it is only a static, 2D graphic representation. Our proposed representation is a pseudo-

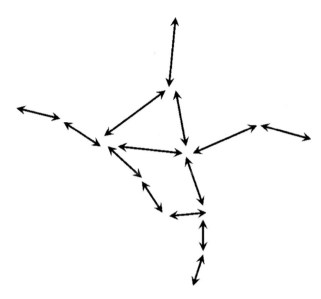

FIGURE 2.4
Visual paths in a photo collage.

3D and interactive representation, the details of which will be introduced in Section 2.3.3.

2.2.2 Hypermedia

Our approach is also based on the hypermedia concept. In computer science, the hypermedia concept was a remarkable discovery. The World Wide Web (WWW), which most of us use daily, is based on hypermedia. The hypermedia structure is very simple. Only two elements, nodes and hyperlinks, are used to create hypermedia. We can apply various kinds of media as nodes, such as texts, images, sounds, movies, and so on.

Historically, hypermedia concept was derived from hypertext concept. Simply saying, hypertext is a content network comprised of texts and hyperlinks (Figures 2.5 and 2.6).

There are several views regarding hypertext (Peter, 2003). One view of hypertext technology is that it is an authoring method. In other words, hypertext is a data-organization paradigm. It augments raw data with a rich, semantically meaningful structure. Hypertext also provides the means to create a context for data. This viewpoint implies that the meaning of data is not inherent in the data itself, but rather resides in the structures into which the data is incorporated. Data is, in itself, only useful when set into a context, structured, and reinterpreted to play a role in a particular setting. Some researchers view hypertext technology as an interaction paradigm; for example, browsing as an alternative to query. We can browse data by continuously

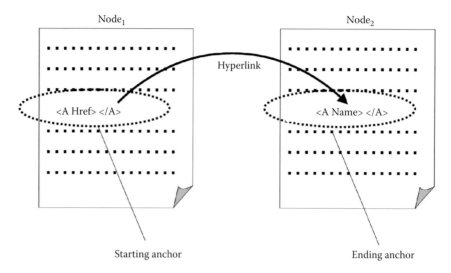

FIGURE 2.5
Hypertest structure based on nodes and hyperlink.

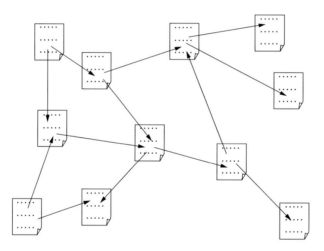

FIGURE 2.6
Hypertext network.

following hyperlinks between nodes. Each path, which typically consists of several hyperlinks, tells us an inherent story.

With the popularization of various media, such as images, sounds, and movies, the hypertext concept has been extended to the hypermedia concept. Today, we can associate various kinds of media, which are not only texts but also images, sounds, and movies, by using hyperlinks. Hypermedia is easy to use for both authoring and browsing, so it has achieved worldwide acceptance and has enabled the remarkable growth of the WWW.

Our STAMP incorporates the hypermedia concept. Our software is based on the *hyperphoto* concept, which can be considered a kind of hypermedia. The details of our hyperphoto concept will be discussed in the following section.

2.3   STAMP: A Tool for Hyperphoto Collage

2.3.1   Concepts and System Overview

Until now, there has been no direct linkage between the photo collage and hypermedia, discussed in Section 2.2. The former is a technique used in the field of art, while the latter is a technique used in the field of computer science. However, we have found that these two techniques are similar in important respects. That is, both are easy to use and enable us to create multiple scenarios from raw data through simple operations. The essential difference between the two techniques is that a photo collage represents a continuous space on a 2D plane, and hypermedia is based on a discrete structure over a computer network. We have tried to merge the advantages of both the photo collage and hypermedia in a concept.

We named our concept *hyperphoto collage*, which is a combination of hyperphoto and photo collage. Hyperphoto means a photo that contains hyperlinks to others. Multiple hyperphotos can form hyperphoto networks, which can be considered a kind of hypermedia.

In our system, users can create spatial associations between raw photos through simple operations. *Spatial-hyperlink*, which is an extension of ordinary hyperlink, is used for creating an association between two photos. We will introduce the details of spatial-hyperlink in Section 2.3.2. Figure 2.7 shows the data structure of hyperphoto networks. The ability to make hyperphoto networks is an authoring technique of our system.

Moreover, our system automatically makes a pseudo-3D photo collage from hyperphoto and displays it (Figure 2.8). This allows us to navigate in pseudo-3D space interactively and continuously. This is our browsing technique. We will discuss this process in detail in Section 2.3.3.

We have implemented two components of STAMP: STAMP-Maker and STAMP-Navigator. STAMP-Maker is an authoring tool, and STAMP-Navigator is a browsing tool. Figure 2.9 illustrates both the flow of editing photos with the two STAMP components and the relations among the three levels of processing data, that is, raw photo, hyperphoto networks, and hyperphoto collage, as visualization of hyperphoto.

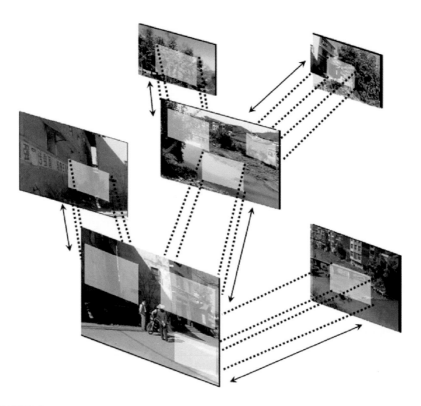

FIGURE 2.7
Hyperphoto networks composed of photos and spatial-hyperlinks.

FIGURE 2.8
Hyperphoto collage automatically made from hyperphoto.

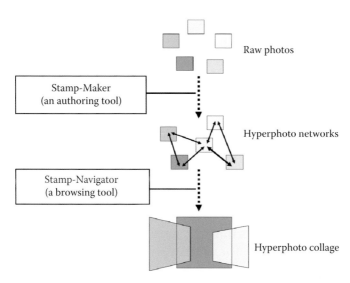

FIGURE 2.9
A flow of editing photos with stamp.

FIGURE 2.10
Creating a spatial-hyperlink between two photos.

2.3.2 STAMP-Maker

STAMP-Maker, one of the STAMP components, is a tool for creating pseudo-3D spaces from multiple photos. The operations to associate photos are very simple. A user specifies the same feature or area in two photos by drawing a rectangle or polygon on each of the two photos. In the example shown in Figure 2.10, the same building appears in two photos, and the user has drawn a rectangle enclosing the building on each of the photos. Specifying two rectangles on two photos attaches a starting anchor and an ending anchor to create a spatial-hyperlink between the two photos. A spatial-hyperlink is comprised of two areal anchors in each photo, which are a starting anchor and an ending anchor. Unlike ordinary hyperlink, spatial-hyperlink enables us to make a relationship between two partial areas in each photo.

Thus, users can create associations between any photos, even photos taken by different people or on different dates. Users can also create additional spatial-hyperlinks on one photo. Therefore, users can create many different routes for spatial navigation (Figure 2.11).

A spatial-hyperlink is basically used to specify the same feature or area in two photos to create an association regarding the spatial meaning in the real world. A more advanced use, however, is to create an association between two photos taken in different places. For example, the same kind of dishes may appear in two photos taken in different restaurants, and a user can create a spatial-hyperlink associating the two photos. In such cases, the photos are linked based on *semantic relationship* over the space rather than a *spatial relationship*. Furthermore, STAMP-Maker also offers advance uses of spatial-hyperlinks, which represent temporal order. In short, these are three types of using spatial-hyperlinks — *spatial relationship*, *semantic relationship*, and *temporal relationship*. Their mechanisms are the same at the system level,

FIGURE 2.11
(See color insert following page 176.) Creating various routes by using spatial-hyperlinks.

and the differences between them exist only at the application level. We will introduce those differences through our example in Section 2.4.

2.3.3   STAMP-Navigator

STAMP-Navigator is a browsing tool that allows people to navigate pseudo-3D spaces through a simple interface. In this component, each spatial-hyperlink is associated with a geometric transformation combining translation, rotation, and distortion. The transformation is defined so that a rectangle specified in the photos at one end of a spatial-hyperlink is transformed into a corresponding rectangle specified in photos at the other end of the spatial-hyperlink. This navigation system displays one focused photo and several linked photos. The focused photo is placed at the center of the display, and linked photos are placed at transformed positions. Projection transformation is used for this process. The focused photo is opaque, while linked photos are translucent and overlaid on the focused photo. When a user selects a photo from the linked photos by directly clicking on it, the scene changes. The navigation system displays a short animation that represents a smooth transition (Figure 2.12). In the transition to the next scene, the currently focused photo fades out and the next focused photo moves to the center. Simultaneously, the transparency values of all photos are changed. This animation gives the user a sense of motion toward the viewpoint of the next focused photo. A photo may be linked to more than one photo, which allows scene transitions to fork into multiple paths. Thus, users can freely step forward, step backward, step sideward, rotate, translate, and zoom in/out in the pseudo-3D spaces by selecting from among candidate photos. This mechanism makes great use of a human's spatial cognition of perspective scenes. STAMP-Navigator thus allows people to experience a sensation similar to walking through the real world.

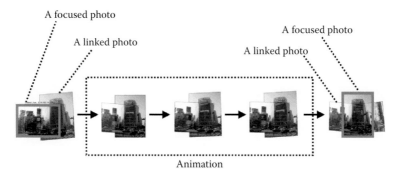

FIGURE 2.12
Smooth transition from one scene to the next.

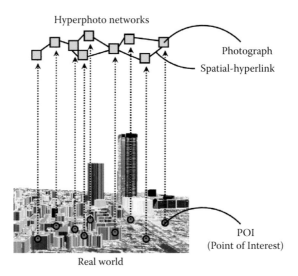

FIGURE 2.13
Conversion from the real world to hyperphoto networks.

2.3.4   Visualization Options

STAMP differs from video in its ability to represent multiple paths in scene transitions, including branches, loops, reverse motions, and terminals. In this sense, STAMP is closer to nonlinear video, but it can be created through a process much easier than nonlinear video editing. An advantage of STAMP is the simplicity it inherits from photography. Also, the structures of hyperphoto networks are very flexible. The structure of hyperphoto networks is, in one sense, indirectly structures of a place in the real world. In other words, hyperphoto networks are the result of an abstraction of the real world (Figure 2.13).

STAMP offers several visualization options of the whole structure of hyperphoto networks. Figure 2.14 shows a temporal map and a spatial map of a hyperphoto network. Each point means hyperphoto, and each line means spatial-hyperlink. By using such maps, we can analyze places and uncover characteristic structures.

2.4   An Application

We have used the STAMP to create several examples of hyperphoto collage. One of us (Konagaya) is a researcher engaging in ethnographic studies and

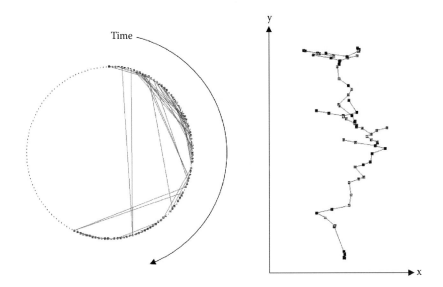

FIGURE 2.14
Temporal map and spatial map of hyperphoto networks.

has taken a large number of photos in Mongolia. We tried to associate these multiple photos with respect to the meaning of their spatial, semantic, and temporal relationships. In ethnographic studies, a research project often consists of three phases — fieldwork, a home stay, and postanalysis. Fieldwork is carried out for the purpose of surveying and understanding the spatial and geographic structure of the target area. After the fieldwork, researchers decide on the area where they will stay. A home stay is the next phase, during which researchers live with native people for about one year. After that, they leave the target area, analyze the data collected, and compile a summary of their studies. Through this process of postanalysis, the researcher can gain a new understanding regarding the target area.

We think these three phases correspond fairly closely to spatial observation, temporal observation, and semantic observation, respectively. Based on these viewpoints, we organized the many photos taken in the Mongolia studies by using spatial-hyperlinks. We can create spatial relationships for photos taken during the fieldwork, temporal relationships for photos taken during the home stay, and semantic relationships for postanalysis.

Figure 2.15 shows an example of the hyperphoto collage we were able to create. We could thus explore the pseudo-3D space and relive the researchers' experiences. In this work, we used semantic relationships to associate photos that contained similar plants and woods. By tracing the three uses of spatial-hyperlinks, we could easily pass through the various points of interest in the target area.

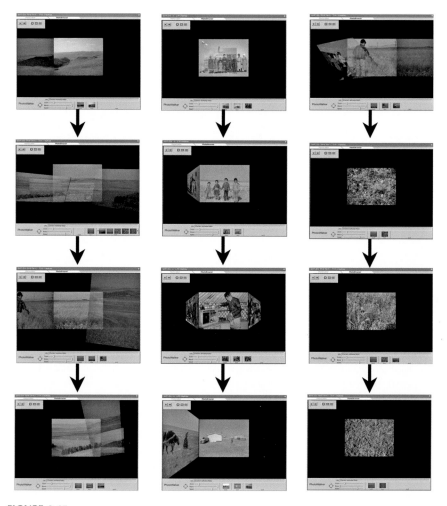

FIGURE 2.15
(See color insert following page 176.) Hyperphoto network created from photos taken in Mongolia — spatial relationship (left), temporal relationship (center), and semantic relationship (right).

2.5 Conclusion

STAMP provides a versatile tool for researchers in the field of ethnographic studies and in other fields, in that it enables them to make greater use of photos in new ways. Although STAMP is not intended as a mathematical or physical analysis tool, it should be useful for presenting information regarding daily activities and experiences. STAMP is a new type of GIS in that it is flexible and enables us to create storytelling/narrative data structures.

Our STAMP software is available for public use. The public-use version of STAMP is named "PhotoWalker." Users can freely download our software from the Web site www.photowalker.net. This Web site also includes additional examples of hyperphoto collages. We believe that STAMP will become widely used in the field of ethnography for various applications and open the new way to the future GIS.

References

Hockney, D. and Alain, S., *David Hockney Photographs*, Rizzoli, 1985.

Peter, J.N., *What Is Hypertext?* International Conference ACM, Hypertext and hypermedia, keynote speech, 2003.

Konagaya, Y., Innovation of Writing on the Mongolian Area Studies, *Geography of the Third World ó from local to global*, Kumagai, K. eds., 1999.

Konagaya, Y., *The Mongolian Nomadic World (CD-ROM)*, TEXNAI, Tokyo, 1998.

Konagaya, Y., *The Spring of Inner Mongolia*, Kawadesyoboshinsha, Tokyo, 1991.

Tanaka, H., Arikawa, M., and Shibasaki, R., *A 3-D Photo Collage System for Spatial Navigations*, Digital Cities II, Computational and Sociological Approaches, lecture notes in computer science 2362, Springer, 2002a, pp. 305–316.

Tanaka, H., Arikawa, M., and Shibasaki, R., *Pseudo-3D Photo-Collage*, International Conference ACM, Web graphics category, Siggraph, conference abstracts, 2002b, p. 317.

Tanaka, H., Arikawa, M., and Shibasaki, R., *World-Wide Gallery for Pseudo-3D Photo-Collage*, International Conference ACM, Siggraph, Web expo category, www.siggraph.org/s2003/conference/web/expo.html, 2003.

3

A Laser-Scanner System for Acquiring Archaeological Data: Case of the Tyre Remains

Ryosuke Shibasaki, Takura Izumi, Hiroya Tanaka,
Masafumi Nakagawa, Yosinori Iwamoto, Hidetomo Fujiwara,
and Dinesh Manandhar

CONTENTS

3.1 Introduction

Excavation in archaeology is conducted to acquire and collect information on archaeological remains and relics in a systematic way using limited time and human resources. Data to be collected are so diversified. The data ranges from overall structure of archaeological remains and relations of strata, details of individual parts of archaeological remains, and information on each relic, such as its classification, location, and strata of unearthed position, its three-dimensional shape, and photos. These voluminous and diversified pieces of information should be efficiently collected, acquired, and organized in such a manner that the relationships among them can be easily retrieved.

In recent years, digital camera, laser scanners, spatial-database management system, such as Geographic Information Systems (GIS), and three-dimensional drawing and modeling tools, such as Computer Aided Design (CAD), have made very rapid progress. The advances make it so easy to acquire digital data on archaeological remains and relics. At the same time, it also provides a possibility of developing new types of products, such as three-dimensional models. In addition, using the Internet, the digital data can be easily shared among archaeologists. Through sharing digital archaeological data among larger numbers of researchers, comparative studies and analysis from more diversified viewpoints can be promoted, which will eventually result in greater contribution to the advances in archaeology.

To actually realize more efficient acquisition and collection of information and sharing in archaeological excavation, how to use and combine advanced sensors, devices, and software has to be discussed and devised. Sensors, data-measurement devices, and software are tools. They require know-how and ideas to effectively apply, just like carpenter tools alone are not enough to build a good house if no skills and know-how are combined with them. Good "design" on how and in which aspects to use, combine them for excavation, and subsequent organizing and analyzing works is really a key. Good design may also reveal some missing links, i.e., a kind of software and devices to be developed especially for archaeological excavation.

This section reports an example of "good design" on how to better use three-dimensional measurement tools, such as laser scanners and data-management tools, such as GIS, including newly developed software and know-how to fill gaps between advanced technologies and the demand in archaeological excavation, through a case study of Tyre remains, Lebanon.

FIGURE 3.1
Laser scanner in Tyre, Lebanon.

3.2 A New, Three-Dimensional Measurement Device: Laser Scanner

For the past four or five years, laser scanners for three-dimensional measurement have become drastically cheaper and smaller, and therefore, much more popular (Figure 3.1). Laser scanners acquire three-dimensional shape data on an object in the following process. At first, as shown in Figure 3.2, a laser scanner emits a laser beam and measures the return time of the beam reflected on the surface of the object. From the travel time of the laser beam, the exact distance between the laser scanner and the object is measured. In parallel, beam angle, i.e., horizontal and vertical angles, are measured. By combining the distance, the horizontal and vertical angle, the three-dimensional coordinate values relative to the laser scanner can be computed. By repeating this process with an incremental change in angles several thousand to several hundred thousand times per second, a very large amount of three-dimensional points are generated. The three-dimensional point data acquired in this manner is called "point-cloud" data. With the three-dimensional point-cloud data, the shape of the object surface is represented. The measurement accuracy usually ranges from several millimeters to several centimeters. Another method of three-dimensional measurement employs photographs and images. A typical example is photogrammetry. By taking pictures of an object from different viewing angles and measuring the location of the object in the photographs or images, three-dimensional location of the object can be estimated. But this measurement process requires exact estimation of position and attitude of a camera or an imaging device in image data acquisition. The estima-

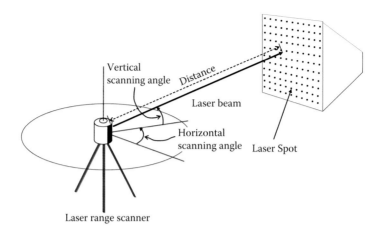

FIGURE 3.2
3D measurement with a ground-based laser scanner.

tion of the position and attitude also require the measurement of the image coordinates of ground control points, or GCPs, with exact ground coordinate values measured in advance. In addition, stereoscopic observation for the three-dimensional measures needs some training. Although cameras themselves have become digital devices that are very easily operated, the preparation and skill needed for three-dimensional measurement make the digital photogrammetry a bit difficult for the ordinary archaeologist. On the other hand, laser scanners, though still quite expensive, make it possible to automate the three-dimensional data acquisition. Automation in measurement is a great advantage of laser scanners over the other measurement devices.

3.3. Architecture of a System for Collecting and Organizing Archaeological Remains Data (Archae-Collector)

3.3.1   Data Types and Objects

Major types of data collected and generated through excavation include drawings, documents, and photos, not limited to three-dimensional measurement data with laser scanners. This chapter proposes the architecture of a system for collecting and organizing data from archaeological excavations, before describing three-dimensional measurement and modeling of archaeological sites. Objects for data collection and generation are classified as follows:

1. Archaeological remains
2. Archaeological relics
3. Excavation work records such as schedule

3.3.1.1   Archaeological Remains

Archaeological remains are mainly represented by a series of drawings, ranging from relatively macroscopic ones of the overall configuration to more microscopic ones on three-dimensional details of individual parts. Drawings reflect the results of judgments on what are important enough to record, as well as the geometric properties of the remains. In this sense, drawings are regarded as a unique form of representation, rather than a symbolic representation of geometric properties. However, because what is considered to be trivial in making drawings may be found to be important afterwards, it is necessary to record source data, such as three-dimensional measurement data and photo data. For the acquisition of three-dimensional data, considering the diversity of archaeological remains in size and required accuracy, the combination of other three-dimensional measurement methods, such as aerial photogrammetry, ground-based photogrammetry, and ground-based survey, rather than laser-scanning devices, should be considered. Moreover, sketches and photos are also important as complementary data to the drawings and three-dimensional measurement data. Especially, photo data can record colors and texture. They can apply to any locations where laser scanners are difficult to apply. In addition, they are effective to let archaeologists easily record with short

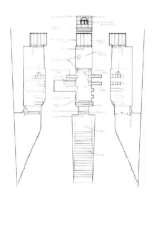

FIGURE 3.3
Example drawings of archaeological remains.

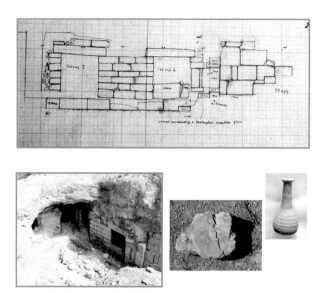

FIGURE 3.4
Examples of handwritten sketches, digital photos of archaeological remains and relics.

memoranda any findings during excavation. In some specific situations, such as excavation of an underground tomb, it may be necessary to record additional sensor data, such as temperature, humidity, and deformation of tomb walls.

3.3.1.2   Archaeological Relics

For recording archaeological relics, drawings, their complementary photos and documents describing classification results, archaeological strata, and estimated era are generated. Three-dimensional geometry of relics can be measured with a laser scanner. For smaller articles, however, stereoscopic measurement with photos and its combination with laser-scanning devices can be applied. Generally, measurement accuracy of laser scanners becomes no better, even though the laser scanners get closer to the objects. On the other hand, photos or images-based measurement, such as photogrammetry, can be more accurate, in case measurement cameras get closer to objects. It can be more advantageous for the measurement of relatively small articles or objects.

3.3.1.3   Other Documents

Information on archaeological relics and remains obtained through an excavation can be easily linked to a daily log of excavation works. These links describe which stages the excavation works are in and what kinds of relics were found in which situation. Daily logs or time records of excavation

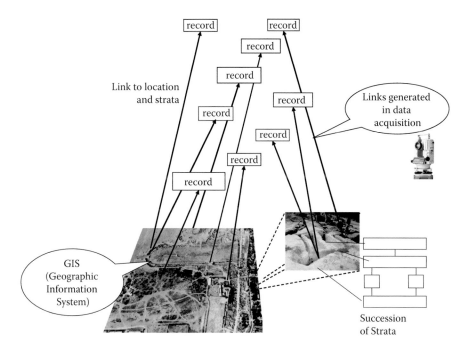

FIGURE 3.5
Links of individual records of archaeological relics and remains to location and strata.

works are regarded as supplementary data in archaeological reports compared with analysis and examination results. However, it is very straightforward and easy to organize information on archaeological relics and remains by connecting them to daily logs or time records of excavation works. Locations can also be easily linked to information on relics for easy data management.

3.3.2 Associations Among Data

To organize a wide variety of data so that users can easily retrieve what they want, it is necessary to provide keys for easy query and to provide associations among data to let users track them in data retrieval. The most fundamental keys are location and time. Unearthed relics and remains can be directly associated with location (Figure 3.5). Time has two classes: time of excavation and strata. Time of excavation can be used to establish links between data or records and daily logs of excavation (Figure 3.6). Establishing links to location and time make it possible to retrieve data from the location and strata where relics and remains are excavated.

In addition, in archaeological reports and articles, information on relics and remains are associated with each other along with the viewpoints of analysis or the context of discussion and speculation. Such associations, if recorded

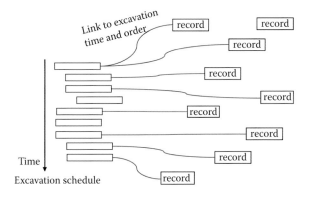

FIGURE 3.6
Links of individual records of archaeological relics and remains to excavation timetables.

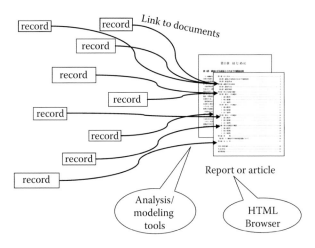

FIGURE 3.7
Links of individual records of archaeological relics and remains to excavation reports and archaeological articles.

explicitly so that other researchers retrieve and examine data along with them, may help others find new aspects or viewpoint of analysis (Figure 3.7).

3.3.3 Architecture of a System for Collecting and Organizing Archaeological Information (Archae-Collector)

To effectively apply devices of collecting and acquiring digital data, such as laser scanners, to archaeological excavations, it is necessary to support the entire process, including the description of associations among various types of collected data, the attachment of space and time keys, and data-management scheme using those keys, as well as to provide measurement methods using laser scanners. Here we name a system for collecting and organizing

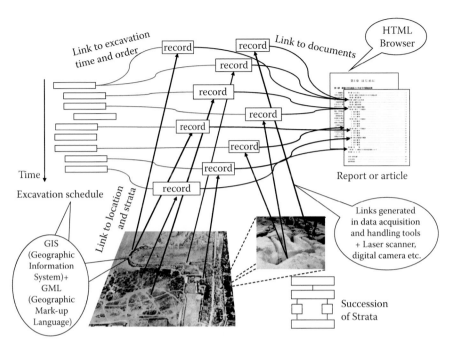

FIGURE 3.8
Supporting, acquiring, and linking archaeological data: An architecture of Archae-Collector.

archaeological information, "Archae-collector," the architecture of which is shown in Figure 3.8. All data records, such as three-dimensional data, drawings, photos, and documents, are linked to space, time, and documents, such as excavation reports or archaeological articles. Tools to help develop and use linked data records are provided so that users make an access to and use necessary data by tracking the links. So far, researches or projects have been conducted on the development of large-scale, three-dimensional models of cultural heritage (e.g., Ikeuchi et al., 2003) and on digital-archive systems for world heritage (e.g., Digital Archive Network for Anthropology and World Heritage [DANA-WH]). The former type of researches focus on the development of three-dimensional models, and the latter types of researches or projects postulate that digital archaeological data are already organized and recorded in a database and that metadata are attached to let users find out necessary information. Archae-Collector focuses on the process ranging from the primary data acquisition by excavation to the establishment of associations or links among the data to enable data management using a database.

Links to space and time can be described mainly by map coordinates and time. In addition to the coordinate systems, geographic or location names and relative position defined within maps and images can be used as spatial tags. Data retrieval through links of space and time can be realized with GIS. Description rules of space and time tags are now being standardized by

International Standardization Organization (ISO) in the form of Geographic Mark-up Language (GML). GML is expected to be International Standard (IS) in one or two years at present (2004). On the other hand, links to data records from reports and articles can be easily implemented with Hyper Text Mark-up Language (HTML).

3.4   Three-Dimensional Data Acquisition and Model Development with a Laser Scanner

The authors applied the combination of different types of laser scanners to the three-dimensional measurement at Tyre remains, Lebanon, because we aim to explore the possibility of laser scanners for archaeological excavation and discuss how laser scanners should be applied and what kinds of software tools and know-how are needed for an archaeologist to use in a variety of excavation works. In the following sections, findings of know-how and development of tools are described based on the Tyre case.

3.4.1   Types of Laser Scanners and Their Combinations

There are trade-off relationships between measurement speed (how many points can be measured per second) and measurement accuracy of a laser scanner. Laser scanners with high scanning speed tend to have lower accuracy, while the improvement of measurement accuracy may lower the measurement speed. Two types of laser scanners can be found in the market, one with higher accuracy with the order of millimeter but with low measurement speed (more than 10 minutes for a single shot), and the other with lower accuracy with the error of several centimeters but with high scanning speed (several tens of seconds to one minutes for a single shot).

Surfaces of an object can be represented more faithfully using three-dimensional measurement with a laser scanner with higher accuracy and resolution. When the surfaces are represented more faithfully with higher density of points, interpolation of the surface and geometric registration of neighboring laser-scanner data can be made more easily, resulting in a higher level of automation. On the other hand, it is not so easy to automate geometric registration and interpolation of laser-scanner data with coarser resolution and lower accuracy.

Some of the laser scanners can measure the reflection intensity of a laser beam from an object. By putting a high reflectivity seal on the surfaces of an object, the seals can be easily recognized in the reflection-intensity images. Those seals can be used as control points for geometric registration of overlapping laser-scanner data and for establishing transformation to a map-coordinate system.

Figure 3.9 summarizes laser scanners and their features applied at Tyre. These are good representatives of different scanner types. Appropriate scan-

Cyrax 2500 (CyraTechnologies /LeicaGeosystems)

Scanning pitch	Horizontal resolution	Vertical resolution
	Min. 0.25mm (50m)	Min. 0.25mm (50m)
	Max.	Max.
	1000points/column	1000points/row
Field of view	Horizontal direction	Vertical direction
	40°	40°

Scanning pitch	Horizontal resolution	Vertical resolution
	0.24°	0.024°
	Max.	Max.
	36000points/column	15000points/row
Field of view	Horizontal direction	Vertical direction
	340°	80°

LMS-Z210 (RIEGL)

Scanning pitch	Horizontal resolution	Vertical resolution
	0.01°	0.018°
	Max.	Max.
	36000points/column	15000points/row
Field of view	Horizontal direction	Vertical direction
	360°	310°

IMAGER5003 (Zoller + Froehlich GmbH)

FIGURE 3.9
Laser scanners applied at Tyre, Lebanon.

ners had to be selected and combined according to the spatial extent of sites and the requirements in measurement accuracy and density of points. In the case of Tyre, LMS-Z210 (Riegl) was mainly applied to a wide Hippodrome, while Cyrax 2500 (Cyra Technology/Leica Geosystems) and IMAGER 5003 (Zoller + Froehlich GmbH) were applied to measure the detail of individual structures and remains. Figure 3.10 shows data examples acquired by the three laser scanners. Cyrax 2500 has relatively high measurement accuracy (several millimeters) and high density of measurement points, though it is not so fast in data acquisition. IMAGER 5003 also has high accuracy and high density and is not so slow in scanning speed. LMS-Z210 is relatively quick in data acquisition and can cover wider areas, because the measurement range is longer than the others, but has less accuracy (several centimeters) and lower density of points compared with the others.

3.4.2   Geometric Registration of Laser-Scanner Data

Since ground-based laser scanners emit a laser beam from a location on the ground, the spatial extent of measurement from that location is limited due to many occlusions. To cover an entire archaeological remain, laser-scanner data obtained at many different locations have to be integrated or registered with each other. As shown in Figures 3.11 and 3.12, at first, corresponding points between neighboring laser-scanner data are to be

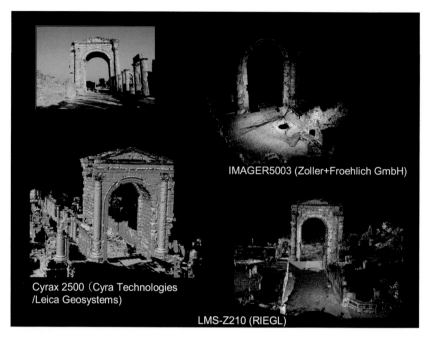

FIGURE 3.10
Data from the three laser scanners.

FIGURE 3.11
Finding corresponding points among neighboring laser-scanner data (Takase et al, 2002).

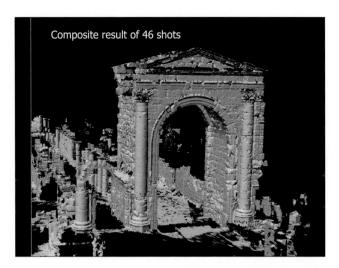

Composite result of 46 shots

FIGURE 3.12
Final results of merging 46 shots (Takase et. al, 2002).

identified. Second, relative position and attitude of the neighboring laser scanners are estimated using the corresponding points. Since Cyrax 2500 data, as shown in Figure 3.10, has relatively high accuracy and higher density of points, it is rather easy to find corresponding points. In the case of LMS-Z210, accuracy and point density are relatively low, and it is not always easy to identify the accurate location of corresponding points. LMS-Z210, however, can record the reflection intensity of reflection points. Small reflection seals on the surface of an object can be easily identified as a bright spot in the reflection-intensity image. By using these bright spots of reflection seals as "tie points," neighboring laser-scanner data can be registered with each other. In addition, by measuring the map-coordinate values of the bright spots, laser-scanner data can be transformed to the map-coordinate system. Figure 3.13 is the final results of the registration of laser data for the Hippodrome.

3.4.3 Reconstruction of Three-Dimensional Shapes from Laser-Scanner Data

Data directly derived from laser scanners are the three-dimensional coordinate values of points that reflect laser beams. Even with a single scan, a large number of data are generated. When the point-cloud data are visualized from viewpoints relatively close to the data, point data on both visible surfaces and occluded surfaces appear on the screen, which may make it hard for users to interpret correctly the three-dimensional geometry among the points. To avoid this, surfaces should be generated through the interpolation of neighboring points. Point data on occluded surfaces are really hidden by the interpolated surfaces, which help a user naturally grasp the

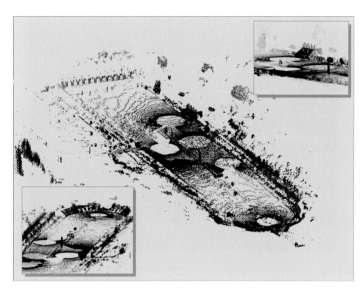

FIGURE 3.13
Results of merging LMS-Z210 data of the Hippodrome.

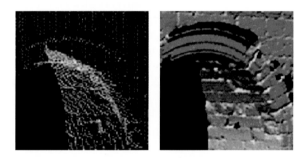

FIGURE 3.14
Comparing point-cloud data with surface patches interpolated from the point-cloud data.

three-dimensional structure (Figure 3.14). In addition, from interpolated surfaces, cross-sectional drawings can be generated by cutting an object with a plane. In interpolating surfaces, usual practice is to select three neighboring points to form a triangle and to define a triangular plane. This network of points connected to generate triangular planes or patches is called Triangulated Irregular Network (TIN).

In interpolating triangular planes, higher-density accuracy of points helps automate a process of finding appropriate neighbors of points and of forming triangular patches. On the other hand, if the accuracy and density is not high enough, it may increase the difficulties in finding out correct neighborhood relationships among points, which may result in the failure of the automated processing. The same kinds of issues may arise, especially in connecting a number of laser-scanner data and in making interpolation of surfaces in places where point-cloud data from several laser scanners are overlapping.

On the other hand, interpolated surfaces from point-cloud data may not represent the geometry of real surfaces very faithfully, if interpolation methods are not correctly chosen. For example, if users apply an interpolation method that tends to excessively smooth sharp peaks, edge lines of the original surfaces may disappear. Selection of surface-interpolation methods has to be made based on geometric properties of original surfaces. It also suggests that raw laser-scanner data (point-cloud data) should be kept together with the interpolated surface data in order to interpolate surfaces again in case it is found that the interpolation results are inappropriate. The original point-cloud data may be helpful even for generating quick-look images, because visualization of surfaces requires a relatively large computational load to evaluate the visibility of individual surfaces from a viewpoint. Point-cloud data require no such computation.

3.4.4 Visualization by Combining Laser-Scanner Data and Digital-Camera Images

Since laser-scanner data measure just the geometric properties of an object, digital-camera images should be "overlaid" to record or visualize colors and texture of the original surfaces. To accurately overlay or drape digital-camera images onto three-dimensional shape data, the relative position and attitude of a digital camera in taking images have to be estimated against the three-dimensional laser-scanner data (Figure 3.15). Estimation of the position and attitude of the digital camera can be made by identifying and measuring the location of corresponding points between digital-camera images and laser-scanner data. A method is proposed to automate the identification of corresponding points. In an example shown in Figure 3.15, a pseudo-digital-camera image, an image which might be taken by a digital camera located at a specific position and viewing angle, is generated using the laser-scanner data and, afterwards, corresponding points are automatically identified between the pseudo-digital-camera image and the real digital-camera image. By using the pseudo-digital-camera image, not the laser-scanner data, directly, the reliability and accuracy of identifying the corresponding points can be improved.

Figure 3.16 shows the results of draping or projecting a digital-camera image onto the laser-scanner data, based on the estimation results of the camera position and attitude. By correctly reconstructing the geometry between the digital camera and the laser-scanner data, we can compute which three-dimensional point in the point-cloud data corresponds to which pixel in the digital-camera data. In this example, colors or red, green, blue data in the digital-camera image are assigned to the corresponding point data in the point-cloud.

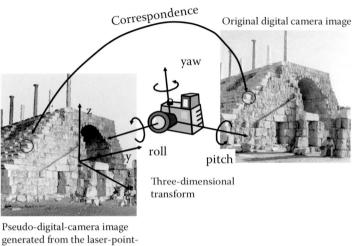

Correspondence

Original digital camera image

yaw

roll

pitch

Three-dimensional transform

Pseudo-digital-camera image generated from the laser-point-cloud data

FIGURE 3.15
Determining the position and attitude of camera against the coordinate system of the laser point-cloud data.

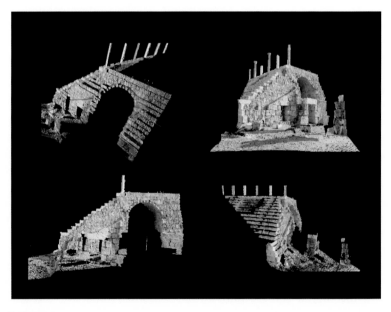

FIGURE 3.16
(See color insert following page 176.) Perspective images of the 3D + textured data of stepped stadium from different viewpoints.

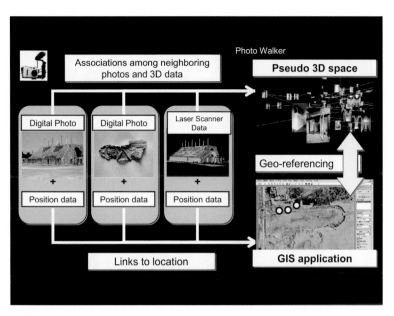

FIGURE 3.17
Outline of links among data generated from Tyre data.

3.5 Implementation of an Example of Archae-Collector

Figure 3.17 is a schematic overview of a prototype of Archae-Collector implemented with PhotoWalker (Tanaka et al., 2001) and GIS. Collected data have not only spatial links but also proximity links representing neighborhood relationships among data. A mechanism that allows users to track and download data along with the proximity links is realized by using PhotoWalker. Figures 3.18 and 3.19 present examples of tracking Tyre data along with the proximity links. Perspective images generated from laser-scanner data are mixed among digital-camera images. The perspective images provide outline information on laser-scanner data so that users can evaluate the relevance of the laser-scanner data before they download large files. PhotoWalker uses Universal Resource Locator (URL) to describe the location or link of the data. If users find it necessary, they can easily move to the FTP site to download point-cloud data from laser-scanner data or the other data products, such as three-dimensional surface data, with colors and texture. In addition, those linked data can be easily made open to public through the Internet.

After users download data products, they may want to make drawings. To support the drawing works, Archae-Collector provides methods and tools. For example, by projecting point-cloud data onto a plane properly selected by a user, users can recognize features of archaeological relics and

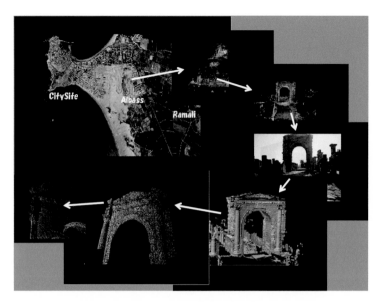

FIGURE 3.18
An example of data links in Archae-Collector (Albaas Area).

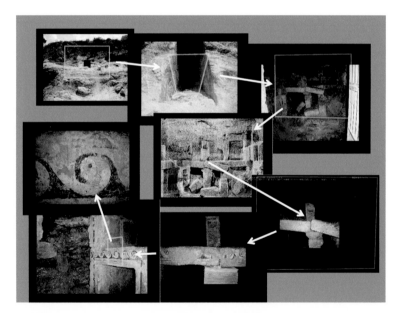

FIGURE 3.19
An example of underground tomb data links in Archae-Collector (Ramali Area).

remains, such as the location of walls and edge lines of stones. Users can draw lines and curves on the projection image. If laser-scanner data are geometrically registered with digital-camera images, a tool is available that

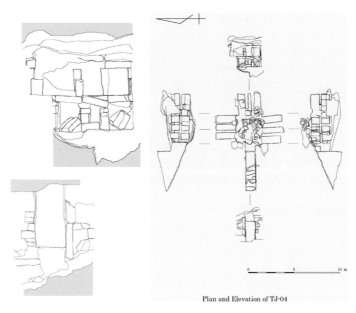

Plan and Elevation of TJ-04

FIGURE 3.20
An example of drawings of the tomb at Ramali, generated from laser data.

allows users to generate three-dimensional lines and curves just by drawing lines and curves directly on the digital-camera images. Behind the tool, when users draw lines and curves on the digital-camera images, three-dimensional laser-point data corresponding to those lines and curves can be extracted. By fitting lines and curves to the extracted three-dimensional points, three-dimensional lines and curves can be generated. By connecting those three-dimensional lines and curves, three-dimensional surface models of relics and remains can be developed. By projecting those lines and curves, drawings can be generated. Figure 3.20 demonstrates example drawings generated with the methods mentioned above.

3.6 Conclusions and Future Prospects

We propose a system called Archae-Collector that includes a method of developing three-dimensional data with colors and texture by geometrically registering and integrating laser-scanner data and digital-camera images and a method of linking those data products using multiple links, such as location, strata, time, and context of analysis. Archae-Collector helps users organize a variety of digital data by establishing links among them without designing complicated data structures for databases. In addition, the data linked with others can be easily made open to the public through the Internet.

With these reasons, we believe Archae-Collector effectively helps archaeologists acquire/collect, organize, and share data among an excavation team, even during the excavation work. In addition, we developed several tools to help make traditional drawings from laser-scanner data and digital-camera imagery. Since this tool requires no skills, such as stereoscopic measurement, archaeologists will find it easy to use for providing traditional excavation reports with drawings and photos. If they find it necessary to conduct a query to the whole datasets, data products developed with Archae-Collector can be transferred to a database, because individual data products already have tags of location, strata, time, etc., and are linked with each other. In this sense, Archae-Collector can be regarded as a quick data-collection and organizing tool to get data well-prepared for the development of a full-fledged database.

Acknowledgment

The authors express special thanks to the members of Nara University Archaeological Team (led by Professor T. Izumi), Yu Fujimoto, Keiji Takase, Susumu Morimoto, and the other team members, Dr. Yutaka Takase and Osamu Yamada (CAD Center Corporation), Masato Shimizu (Kokusai Kogyo Corporation), Ryutaro Okugawa, Akira Iwata, and Hiroyasu Sasaki (Toshiba Engineering Corporation) for their contributions.

References

Digital Archive Network for Anthropology and World Heritage (DANA-WH), www.dana-wh.net/home/.

Fujiwara, H., Nakagawa, M., and Shibasaki, R., Automated texture mapping for 3D modeling of objects with complex shapes — a case study of archaeological remains, The 24th Asian Conference on Remote Sensing, Busan, Korea, 2003.

Ikeuchi, K., Nakazawa, A., Hasegawa, K., and Ohishi, T., *The Great Buddha Project: Modeling Cultural Heritage for VR Systems through Observation*, IEEE ISMAR03, Tokyo, 2003.

Takase, Y., Sasaki, Y., Nakagawa, M., Shimizu, M., Yamada, O., Izumi, T., and Shibasaki, R., *Reconstruction With Laser Scanning and 3D Visualization of Roman Monuments and Remains in Tyre, Lebanon*, proceedings of ISPRS WG V/4 and IC WGIII/V, (CD-ROM), 2002.

Tanaka, H., Arikawa, M., and Shibazaki, R., A 3D photo collage system for spatial navigations, International Conference Digital City Workshops, 2001.

4

A Laser-Scanner System for Acquiring Walking-Trajectory Data and Its Possible Application to Behavioral Science

Huijing Zhao, Katsuyuki Nakamura, and Ryosuke Shibasaki

CONTENTS

4.1 Introduction

Monitoring and analyzing human movement, such as tracing pedestrians in a crowded station plaza and analyzing their walking behavior, is considered to be very important in behavioral science, sociology, environmental psychology, and human engineering. So far, motion analysis using video data has been the major method to collect such data. A good survey of visual-based surveillance can be found in Gavrila (1999). The following are several examples that target tracking a relatively large crowd in an open area. Regaz-zoni and Tesei (1996) described a video-based system for counting people over a period of time and detecting overcrowded situations in underground railway stations. Schofield et al. (1997) developed a lift-aiding system by counting the number of passengers waiting at each floor. Uchida et al. (2000) tracked pedestrians on a street. Sacchi et al. (2001) proposed a monitoring application, where crowds moving in an outdoor tourist site were counted using a video image, and Pai et al. (2004) proposed a system of detecting and tracking pedestrians at crossroads to prevent traffic accidents.

One of the difficulties of using video cameras is that they do not cover the entire viewing area, and out-of-sight areas, called occlusions, exist. Image resolution and viewing angles are limited due to such "camera settings" so that a moving object that has fewer image pixels may fail to be tracked. Unceasing changes in illumination and the weather are another major obstacle affecting the reliability and robustness of a visual-based system. In order to cover a large area, multiple cameras are used. However, the data from different cameras can be difficult to combine, especially in a real-time process, as this requires accurate calibration and complicated calculations to account for the different perspective coordinate systems. Up until now, the application of visual-based surveillance has been limited to the extraction of a few objects in rather limited environments.

Recently, a new sensor technology, single-row laser (range) scanners, has appeared. It profiles across a plane using a laser that is nonharmful to the human eye (Class 1A laser, operating in the near-infrared part of the spectrum). This measures the distance to a target object according to, for example, the time of flight at each controlled beam direction. In recent years, single-row laser (range) scanners (hereafter "laser scanner") having a high scanning rate, wide viewing angle, and long range have been developed and can be acquired commercially at cheap prices. These have attracted increasing attention in the field of moving-object detection and tracking. Applications can be found in Streller et al. (2002), where a laser scanner was located on a car to monitor a traffic scene; in Prassler et al. (1999), where a laser scanner was set on a wheelchair to track surrounding people to help a handicapped person travel through a crowded environment, such as a railway station during rush hour; and in Fod et al. (2002), where a laser-based, people-tracking system is presented.

In this research, we propose a novel tracking system aimed at providing real-time monitoring of pedestrian behaviors in a crowded environment, such as a railway station, shopping mall, or exhibition hall. A number of single-row laser scanners are used to cover a large area to reduce occlusions. The distributed data from different laser scanners are spatially and temporally integrated into a global-coordinate system in real time. A pedestrian-walking model was defined, and a tracking method utilizing a Kalman filter (for example, Jang et al., 1997; Sacchi et al., 2001; and Welch and Bishop, 2001) was developed. The major difference between our system and that of Fod et al. (2002) is that Fod et al. (2002) set their laser scanners to target the waist height of an average walking person. In contrast, we place our laser scanners at ground level to monitor pedestrians' feet and track the rhythmic pattern of walking feet. There are several reasons: The occlusion at ground level is much lower than at waist height; the reflections occur from swinging arms, hand bags, and coats are difficult to model to obtain an accurate tracking; and the rhythmic, swinging feet are the common pattern for a normal pedestrian, which can be measured at the same horizontal plane.

In the following sections, Section 4.2 outlines the sensor system, data acquisition, moving-object extraction, and distributed data integrations. Section 4.3 defines a pedestrian-walking model, followed by an explanation of the Kalman filter-based tracking algorithm. Section 4.4 evaluates the system using an all-day experiment conducted at a railway station. The pedestrian flow was analyzed spatially and temporally, suggesting a possible application of the technique to behavioral studies.

4.2   Outline of the System

4.2.1   Single-Row Laser Scanner and Moving-Object Extraction

Two types of single-row laser scanners have been studied in this research, LMS200 by SICK and LDA by IBEO Lasertechnik (Figure 4.1). Here, we introduce a sensor's specification and configuration using the LMS200 as an example. When scanning within an angle of 180° at a resolution of 0.5°, a scanning rate of about 37 Hz is reached. In each scan, 361 range values are equally sampled on the scanning plane, within a maximum distance of 30 m, with an average range error of about 3 cm. Both the maximum distance and the average range error vary with the material of a target object. Range values can be easily converted into rectangular coordinates (laser points) using the controlled angle of each laser beam. The coordinates here are in respect to the local coordinate system of the laser scanner. In this research, the laser scanners are set on the floor to perform horizontal scanning, so that cross-sections at the same horizontal level containing data from moving objects (e.g., feet) and motionless objects (e.g., building walls, desks, chairs, and so on) were obtained in a rectangular coordinate system of real dimension.

A background image containing only the motionless objects is generated and updated at each time interval (e.g., every 30 min) as follows. For each beam direction, a histogram is generated using the range values measured in the direction of all laser scans. If a pick above a certain critical value is found out, which denotes that an object is continuously measured in the direction at the distance, it is defined as a motionless object. The background image is composed of the pick values for all the beam directions. The number

FIGURE 4.1
A single-row laser (range) scanner at an experimental site.

of laser scans used in background-image generation and the time interval for background-image updating are set on a case-by-case basis, according to the environment being measured. In the case where the physical layout of the environment does not change often (e.g., an exhibition hall and a railway station), a background image is generated previously and not updated on the air to avoid mishandling of the range values.

Whenever a new laser scan is recorded, background subtraction is conducted at the level of each beam direction. If the difference between two range values is larger than a given threshold (considering the fluctuations in range measurement), the newly measured range value is extracted as data of a moving object. Figure 4.2 shows a sample laser scan, where the laser points are classified using background subtraction and shown at different intensities.

4.2.2   Integration of Multiple Single-Row Laser Scanners

A number of laser scanners are exploited so that a relatively large area can be covered, while occlusions and crossing problems can be solved to some extent. Each laser scanner is located at a separate position and controlled by a client computer. All the client computers are connected through a local area network (LAN) to a server computer, which gathers the laser points of all the moving objects from all the client computers and conducts the tracking mission.

Since laser points are recorded by each laser scanner at its local coordinate system using the client computer's local time, they are integrated into a global coordinate system before being processed for tracking, where integration is conducted in spatial (x- and y-axis) and temporal (time-axis) levels.

The locations of the laser scanners need to be carefully planned. All the laser scanners form an interconnected network, and the laser scans between each pair of neighboring laser scanners maintain a certain degree of overlap. The relative transformations between the local coordinate systems of a pair

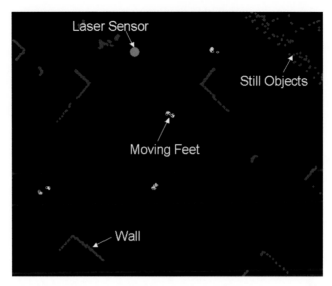

FIGURE 4.2
A sample laser scan. The laser points are classified using background subtraction and shown at different intensities.

of neighboring laser scanners are calculated by pair, wisely matching their background images using the measurements to common objects. In specifying a given sensor's local coordinate system as the global coordinate system, the laser points from each laser scanner can be transformed into the global coordinate system by sequentially aligning the relative transformations. Details on registering multiple laser scanners can be found in Zhao and Shibasaki (2001).

4.3 Tracking Algorithm

A tracking algorithm was developed assuming that the moving objects are solely the feet of normal pedestrians only. In this section, the flow of the tracking process is introduced first to provide a global view of the algorithm. A tracking algorithm utilizing a Kalman filter is then discussed, where a pedestrian-walking model is defined based on the rhythmic swing of pedestrian feet.

4.3.1 Flow of the Tracking Process

A tracking algorithm is designed, as shown in Figure 4.3. In each iteration, the server computer gathers the laser points of moving feet ("moving point")

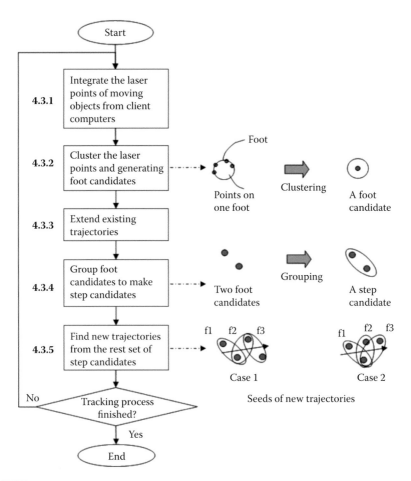

FIGURE 4.3
A flowchart of the tracking process.

in the latest laser scans from all the client computers and integrates them into the global coordinate system to make a frame (Step 4.3.1). Since there may be many points impinging upon the same foot, where the number of points and their spatial resolution relate to the distance from the pedestrian to the laser scanner, a process is initially conducted to the integrated frame to cluster the moving points within a radius less than a normal foot (e.g., 15 cm). The center points of the clusters are treated as foot candidates (Step 4.3.2). Trajectory tracking is conducted by first extending the trajectories that have been extracted in previous frames, then looking for the seeds of new trajectories from the foot candidates that are not associated with any existing trajectories.

A tracing algorithm utilizing Kalman filter is developed to extend the existing trajectories to the current frame (Step 4.3.3). This will be addressed in detail in a later section. The seeds of the new trajectories are extracted in two steps. The foot candidates that are not associated with any trajectory are first paired into step candidates (pedestrian candidates) if the Euclidean distance between them is less than a normal step size (e.g., 50 cm) (Step 4.3.4). A foot candidate could belong to a number of step candidates, if there were multiple options. A seed trajectory is then extracted along more than three of the previous frames, which satisfies the following two conditions. The first is that the two-step candidates in successive frames should overlap at the position of at least one-foot candidate. Second, the motion vector decided by the other pair of nonoverlapping foot candidates should change smoothly along the frame sequence (Step 4.3.5).

4.3.2 Definition of the Pedestrian-Walking Model

When a normal pedestrian steps forward, a typical characteristic is that at any moment, one foot swings by, pivoting on the other foot. The two feet interchange in the step by landing and then shifting in a rhythmic pattern.

According to the ballistic walking model proposed by Mochon and McMahon (1980), muscles act only to establish an initial position and velocity of the feet at the beginning half of the swing phase, then remain inactive throughout the rest half of the swing phase. Here the initial position refers to the situation where a swing foot and a stance foot meet together. In this research, we consider the position, speed, and acceleration of the feet in a horizontal plane, the values of which are in respect to the two-dimensional global coordinate system addressed in the previous sections. In the case the speed of the left foot is faster than the speed of the right foot, the left foot swings forward by pivoting on the right foot. At the beginning half of the swing phase, the left foot shifts from the rear to the initial position, and swings from standing still to an accelerated speed. Here, the acceleration is a function of the muscle's strength. During the remaining half of the swing phase, the left foot shifts from the initial position to the front, and swings with a decelerated speed from a certain speed to standing still. Here, the acceleration is opposite to the walking direction, which arises from forces other than those from left-foot muscles. During the entire swing phase, the right foot remains almost stationary, so that the speed and acceleration on the right foot are almost zero. In the same way, we can deduce the speed and acceleration parameters when the right foot swings forward by pivoting on the left foot. In this research, we simplify the pedestrian-walking model by assuming that the acceleration and deceleration on both feet from either the muscles or from other forces are equal and constant during each swing phase, and they experience only smooth changes as the pedestrian steps forward. Figure 4.4 shows an example of the simplified-pedestrian walking model.

4.3.3 Definition of the State Model

As has been described in the previous section, the pedestrian walking model consists of three types of state parameters: position, speed, and acceleration.

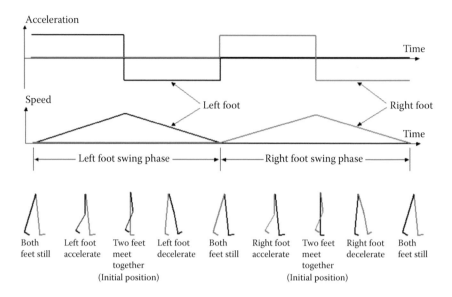

FIGURE 4.4
An example of a simplified pedestrian-walking model.

The position and speed change with acceleration, while the acceleration changes with the swing phase. A discrete Kalman filter is designed in this research by dividing the state parameters into two vectors as follows:

$$s_k = \Phi \; s_{k-1} + \Psi \; u_k + \omega \; , \tag{4.1}$$

where, s_k is a vector, (position, speed) of both feet of a pedestrian at frame k, while u_k is a vector (parameter for position, parameter for speed) of the acceleration. The term ω is a vector (error for position, error for speed) of the state estimation. The transition matrix Φ relates the (position, speed) vector at a previous time step, k-1, to that of the current time step, k, while Ψ relates the acceleration (parameter for position, parameter for speed) vector to the change in the (position, speed) vector.

The discrete Kalman filter updates the state vector of s_k based on the measurements as follows:

$$m_k = H \, s_k + \varepsilon \tag{4.2}$$

where m_k denotes the measured (position, speed) vector, i.e., the position and speed vector calculated from the laser points at time step k. The term H relates the (position, speed) vector, s_k, to the measured (position, speed) vector, m_k, and the term ε denotes the error vector resulting from the measurement.

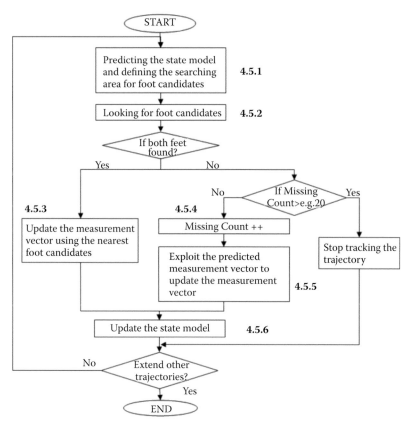

FIGURE 4.5
A flowchart of extending existing trajectories using a Kalman filter.

4.3.4 The Tracing Process Using the Kalman Filter

Figure 4.5 shows the flow of extending the existing trajectories to the current frame. In the extending of each trajectory, the state vector, $u_{k,n}$, is first predicted by identifying the swing phase, and $s_{k,n}$ and $m_{k,n}$ are then predicted using Equations 1 and 2, respectively (Step 4.5.1). A searching area is defined on the predicted $m_{k,n}$ (Step 4.5.2). If any foot candidates of the current frame are found inside the search area, the nearest foot candidates are exploited to compose an updated $m_{k,n}$ (Step 4.5.3). Otherwise, the missing counter starts (Step 4.5.4). If the missing counter is larger than a given threshold, e.g., 20 frames (≈ 2 sec), then the tracing of the trajectory stops. Otherwise, the predicted $m_{k,n}$ is exploited as an updated value (Step 4.5.5) to update the state vector, $s_{k,n}$, and Kalman gain (Step 4.5.6). This process continues until all the trajectories are traced.

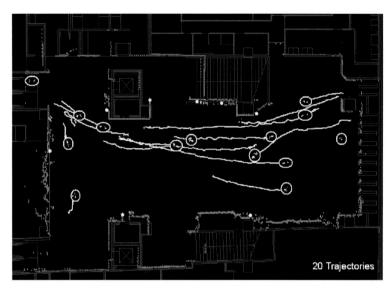

FIGURE 4.6
An example of the reproduction of pedestrian trajectories at a concourse.

4.4 Possible Applications to Behavioral Science

An experiment was conducted in a railway station by monitoring passenger behavior in the concourse over a whole day. The size of the concourse was about 30 × 20 m². During the rush hour, more than 100 passengers occupy the concourse simultaneously. Eight SICK LMS200s were used to cover the concourse, as shown in Figure 4.6, where their locations are denoted by opaque, white circles. Each SICK LMS200 was controlled using a notebook computer (the client computer) with a central processor unit (CPU) speed of more than 600 MHz. These were connected to a server computer using a 10/100 Base LAN cable. The background images were generated by the client computers in the early morning, when the number of passengers inside the concourse was low. These were not refreshed during the data-acquisition measurements. A server computer with a CPU speed of 1 GHz was able to perform a real-time tracking of up to 30 trajectories simultaneously. Since there were many more passengers in the concourse in this experiment, especially during rush hour, passenger trajectories were extracted through a postprocessing.

Figure 4.6 shows an example of the reproduction of pedestrian trajectories inside the concourse. The bright-gray points are the laser points belonging to the background images, the white points are the laser points of moving feet, the transparent circles group the laser points of one person, and the lines represent the trajectories in the latest 50 frames. A dark-gray map has been overlapped in Figure 4.6 to provide a better visualization and understanding of the surroundings. The experimental data were processed on two levels: (1) to assess the reliability of the system and (2) to analyze the change in pedestrian flow.

4.4.1 Assessment of the System Reliability

The following questions are always asked: "What percentage of pedestrians is measured, especially during rush hour?" and "How does it change with time and influence on tracking performance?" Now, let us answer these questions. If a pedestrian is inside the laser scanners' measurement range but cannot be measured, then an occlusion occurs. The laser beams may be blocked either by motionless objects, e.g., building walls, chairs, and desks, or by moving objects, e.g., pedestrians. The occlusions arising from motionless objects do not change with time, so that can be predicted and, to some extent, reduced by arranging laser scanners' locations. On the other hand, the occlusions caused by moving objects change dramatically with time and strongly influence the tracking performance. In particular, if a pedestrian is blocked for a short period, e.g., less than 10 frames, then their trajectory may be predicted using history data. If a pedestrian is continuously blocked, e.g., for more than 20 frames, then their trajectory will be broken. This was addressed in the previous section. Here, we analyze the spatial distribution and temporal change in the occlusions, using a map called an "occlusion map." We analyze the reason of occlusions, as well as their continuity, using a value called the "occlusion ratio."

We tessellated the concourse into grid pixels. An occlusion map was generated by assigning the pixel values to the number of laser scanners able to measure the center of a grid pixel at a given moment or period. If a number of frames are examined to determine whether a grid pixel is continuously blocked, then the average number of visible laser scanners is assigned to the pixel value. Figure 4.7 shows an occlusion map formed at 7 p.m. (before the rush hour) and at 8:30 p.m. (in the rush hour). The bright gray denotes a

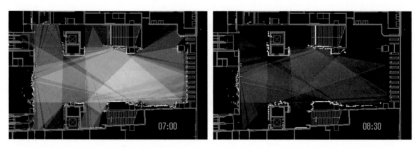

FIGURE 4.7
An assessment of the occlusions from pedestrians.

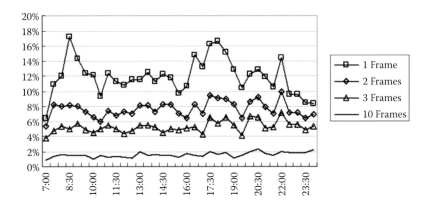

FIGURE 4.8
The change in occlusion ratio with time and with the number of continuous frames.

clear view, the dark gray denotes a poor view, and the black denotes a totally blocked view. Although the occlusion map taken at 8:30 p.m. is much darker compared to the occlusion map taken at 7 p.m., most of the grid pixels inside of the concourse are not black, meaning that the grid pixel can be measured by at least one laser scanner. The occlusion ratio was calculated for each occlusion map using the number of pixels that were blocked by other moving objects (passengers) as the numerator, using the number of pixels that were not blocked by other motionless (background) objects as the denominator. Figure 4.8 shows the change in occlusion ratio with time, as well as with the number of continuous frames. It can be seen that the occlusion ratio was high for single frame, whereas less than 2 percent of the grid pixels were continuously blocked by moving objects (other pedestrians) over a period 10 frames (< 0.5 sec).

On the other hand, an examination was conducted using video images as the ground reference to determine whether, and to what percentage, the pedestrians were tracked accurately. The laser points, as well as the tracking results, were back-projected onto the video images through calibration. Erroneous and lost trajectories were counted using a manual operation and evaluated with respect to the change in pedestrian spatial density. Evaluation of the results showed that almost a 100 percent tracking accuracy was achieved for a spatial density less than 0.4 persons/ m^2. Figure 4.9 shows a back-projection for a spatial density about 0.38 persons/ m^2.

4.4.2 Analyzing the Pedestrian Flow

Our experiments lasted from early morning until late night in a working day. By analyzing the laser points of moving objects and the pedestrian trajectories, the passenger flow inside the concourse, as well as its change with time, can be easily determined. Figure 4.10 shows the change in passenger numbers deduced by counting the pedestrian trajectories, where the

FIGURE 4.9
A back-projection of laser points onto a video image, where the spatial resolution was about 0.38 person/ m^2.

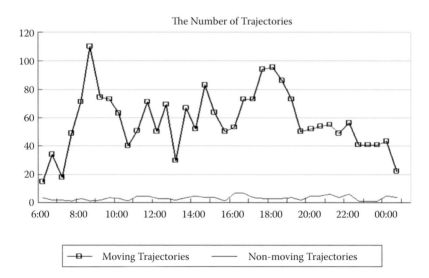

FIGURE 4.10
The number of pedestrian trajectories and their change with time.

trajectories were counted as being either moving ones or nonmoving trajectories (e.g., moving at a speed less that 0.3 m/s). Figure 4.11 shows the distribution and density map of passengers at 7 p.m. (before the rush hour) and 8:30 p.m. (in the rush hour). The dark gray denotes a low, nonzero

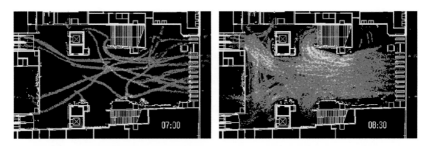

FIGURE 4.11
Passenger distribution and density map.

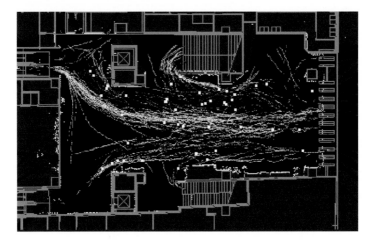

FIGURE 4.12
Oriented flow lines and collision distribution.

passenger density, while the bright gray denotes a high passenger density. Figure 4.12 shows the oriented flow lines and collision distribution, where the bright lines denote people moving to the right, and the dark lines denote an opposite flow lines. In Figure 4.12, the white points show collision points where two passengers get close to each other, within 60 cm.

4.5   Conclusion

A novel method has been proposed to track pedestrians in wide, open areas, such as shopping malls and exhibition halls, using a number of single-row laser (range) scanners. The system was examined through a one-day experiment at a railway station, where, during rush hour, more than 100 trajectories were counted simultaneously. The passenger flow, as well as its change with time, was examined, the result of which might be applied to the microscale behavioral study. Although the tracking algorithm is still not robust and accurate enough to follow each individual and track the complete trajectories of a large crowd, the efficiency of our system in examining pedestrian flow and determining its tendency in a wide and open area has been shown. Compared with the tracking using normal video cameras, it can be concluded that our method of using laser scanners has the following advantages. First, it is a form of direct measurement. The extraction of moving objects in a real-world coordinate system is not as time-consuming a task as using a normal video camera. Second, as the range measurement can be converted into a rectangular coordinate system with a real dimension on a horizontal plane, it is comparatively easy to calibrate multiple laser scanners and integrate the distributed data to cover a relatively large area. Third, the tracking of a large crowd will be achieved in real time in the near future due to the low computation cost. Finally, although range measurements have poor interpretability compared with video images, to some extent this avoids a privacy problem, which is a sensitive topic in public places, such as supermarkets and exhibition halls.

In future work, a tracking algorithm will be developed for the monitoring of an environment of not only pedestrians, but also shopping carts, baby cars, bicycles, motor cars, and so on.

Acknowledgments

We would like to express our appreciation to Kiyoshi Sakamoto from the East Japan Railway Co., Tomowo Ooga from the Asia Air Survey Co. Ltd., and to Naoki Suzukawa from JR East Consultant. They cooperated in the experiments carried out at the railway station and assisted in data processing, and their guidance in flow analysis enabled this research to be a success.

References

Fod, A., Howard, A., and Mataric´, M.J., Laser-based people tracking, Proc. of the IEEE International Conference on Robotics and Automation (ICRA-02), Washington D.C., 2002, pp. 3024–3029.

Gavrila, D., The visual analysis of human movement: a survey, *Comput. Vision Image Understand.*, 73(1) 82–98, 1999.

Jang, D.S., Kim, G.Y., and Choi, H.I., Model-based tracking of moving object, *Pattern Recog.*, 30(6), 999–1008, 1997.

Mochon, S. and McMahon, T.A., Ballistic walking, *J. Biomechanics*, 13, 49–57, 1980.

Pai, C.J., et al., (2004) Pedestrian detection and tracking at crossroads, *Pat. Recog*, vol 37. no 5. pp. 1025–1034, 2004.

Prassler, E., Scholz, J., and Elfes, A., (1999) Tracking People in a Railway Station During Rush-Hour, in *Proc. ICVS*. H.J. Christensen Ed., 1999, pp 162–179.

Regazzoni, C.S. and Tesei, A., Distributed data fusion for real-time crowding estimation, *Sig. Process.*, 53, 47–63, 1996.

Sacchi, C., et al., Advanced image-processing tools for counting people in tourist site-monitoring applications, *Sig. Process.*, 81, 1017–1040, 2001.

Schofield, T.J., Stonham, A.J., and Mehta, P.A., Automated people counting to aid lift control, *Autom. Constr.*, 6, 437–445, 1997.

Streller, D., Furstenberg, K., and Dietmayer, K.C.J., Vehicle and object models for robust tracking in traffic scenes using laser range images, IEEE 5th International Conference on Intelligent Transportation System, September 2002, pp. 118–123.

Uchida, K., Miura, J., and Shirai, Y., Tracking multiple pedestrians in crowd, IAPR workshop on MVA, November 2000, pp. 533–536.

Welch, G. and Bishop, G., An introduction to the Kalman filter, UNC-Chapel Hill, TR95-041, February 2001.

Zhao, H. and Shibasaki, R. (2001) A robust method for registering ground-based laser range images of urban outdoor environment, *Photogram. Eng. Remote Sensing*, 67, 1143–1153.

5

A Method for Constructing a Historical Population-Grid Database from Old Maps and Its Applications

Yoshio Arai and Shiro Koike

CONTENTS

5.1   Introduction: Can GIS Deal with Historical Phenomena?

Owing to recent progress made in the refinement of geographic information systems (GIS), spatial analysis using GIS is penetrating human and social sciences, such as economics, sociology, archaeology, and human geography. However, many studies using GIS in these fields are concerned with relatively recent phenomena that have occurred in the last two or three decades. Few studies deal with long-term events, such as urban growth during the social modernization process of the past 100 years.

Although it is considered that a detailed spatio-temporal analysis of the long-term development of urban areas provides some valuable insights into the nature of cities, severe difficulties are encountered using GIS to study this process. The largest problem is the lack of suitable detailed historical spatial data.

Can GIS really deal with historical phenomena? The only way to make it possible is by digitizing contemporary maps and documents. A few studies have attempted to adapt historical data to be suitable for GIS. Pioneering work in this field was undertaken by Norton (1976), who reconstructed land-use data for a township in Canada. A recent example of research by Lee (1996) is a quantitative analysis of population distribution in Northern Ireland during the 19th century. More recently, Taniuchi (1995) estimated the population-grid data for Tokyo around the year 1900, using various statistical materials. Siebert (2000) reconstructed with GIS the infrastructure patterns of pre–World War II Tokyo using old maps. These studies, however, have a limitation in that they generally focused upon a small area at one point in time and did not cover a much larger area over a long period.

In this chapter, we propose a method for constructing *historical population-grid data* (HPD) from old topographical maps. The method is designed to overcome limitations found in previous studies. Topographical maps, which were made in accordance with an authorized format for survey publications, illustrate geographical features across a wide area. From them we can see the detail of urban and rural areas at the time of map production. Can we derive from these old maps numerical data on the population distribution in those days?

Herein, we have attempted to assemble historical population-grid data from around 1890 (1890-HPD) and that from about 1930 (1930-HPD). These data, together with modern statistics, have been used to make a time-series dataset for intervals of less than 40 years, covering the 110-year period from the close of the 19th century. This dataset is valuable for spatio-temporal analysis of the changing pattern of population throughout the period from the beginning of the Japanese modern era to the present day.

The chapter consists of six sections, including this introductory Section 5.1. A preliminary study in Section 5.2 gives a systematic sampling method for estimating population from an area occupied by housing. By modifying this method, the 1890-HPD is made for the Kanto Plain area in Section 5.3. Section 5.4 constructs the 1930-HPD in the same area, but the manner of derivation is further modified, because the source materials were different. Section 5.5 integrates the 1890-HPD and the 1930-HPD with the population-grid data published by the Statistical Bureau of Japan for 1970 and 2000. Section 5.5 also shows two applications of the integrated 1890-2000 HPD to population studies. The chapter ends in Section 5.6 with suggestions for further work.

5.2 A Preliminary Study on the Population Estimation Made for Around the Year 1890 in East Biwa

To derive the historical population-grid data (HPD) for the Kanto Plain between the years 1890–2000, we carried out the preliminary study for East Biwa shown in this section.

5.2.1 Estimation Method

The means of estimating population from topographical maps published in 1890 is based on the method proposed by Arai and Koike (2003) and Koike and Arai (2001). It is assumed that in 1890 almost all the buildings in villages shown on the maps were single story, and that the family structure was not very varied among settlements. In consequence, the total area occupied by buildings and the number of residents in a village were closely related. From these assumptions, we derived the hypothesis that population P can be expressed as a function of the area of buildings A, appearing on topographical maps, i.e.,

$$P = f(A) \tag{5.1}$$

To measure the area of buildings efficiently, the systematic point-sampling method was used. We overlaid a scale grid of 20-meter intervals on a map and regarded the lattice points as systematic sample points. The number of such points included in the area of buildings was counted (Figure 5.1). Let N be the number of sample points included in the area of buildings. As N is proportional to A, A is substituted by N and Equation (5.1) is written as

$$P = f(N) \tag{5.2}$$

Let us see if the relationship of Equation (5.2) really holds, using empirical data. The study area is fully named the "East-*Biwako* Area" located in the central part of Honshu, the main island of Japan. Old *Seishikizu* topographical

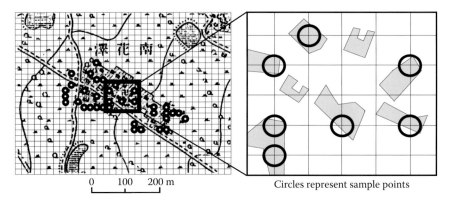

FIGURE 5.1
An example of point-sampling.

maps cover the whole area at a scale of 1/20,000, and these show the geographical features existing in 1890. *Shigaken Bussanshi*, a statistical report, was made at almost the same time as the maps. Actual population numbers by village are given in this report.

We related the number of sample points included in the area of buildings to the population data. The relationship between the actual population of each village *P*, and the number of sample points within the mapped buildings in that village *N*, is shown in Figure 5.2. This figure shows a clear correlation between *N* and *P*. A regression analysis provided the following equation:

$$log(P) = 1.3008N^{0.1634}$$
(5.3)

The fact that the squared correlation coefficient for this regression equation was 0.8304 is positive evidence that there is a close relationship between *P* and *N*, represented by Equation (5.2).

The spatial unit used in Equation (5.3) was a village area, but any spatial unit can be used. In the HPD in East Biwa, square 1 km² cells are used, and these are also employed in the *Basic Grid-Square* (BGS) system, which is now extensively used for GIS-based spatial analysis in Japan.

5.2.2   Estimation Accuracy

The accuracy of the population estimation made above is next examined. We obtained the actual population of each square cell directly from the historical report. As a result, the estimated and actual population of each cell was available for comparison. Figure 5.3 shows the distribution of errors. The percentage of cells having an error within 20 percent was more than 60 percent of the total.

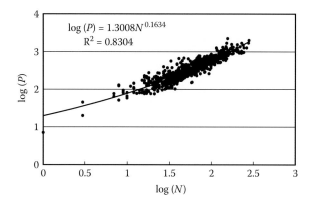

FIGURE 5.2
Relationship of N and P in logarithmic scale (taken from Arai and Koike, 2003).

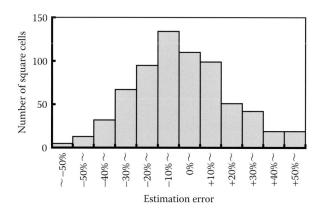

FIGURE 5.3
Distribution of the error rate in the estimated grid population (taken from Arai and Koike, 2003).

An uneven geographical distribution of error was found. The error rate seemed to differ between individual map sheets. This unevenness was caused by an arbitrary cartographic representation of villages or of buildings, resulting from a lack of standardization in map ornament at the time of map drafting.

We attempted to reduce the uneven errors by considering the statistical characteristics of the Chinese lettering size used for village names written on the maps, and this adjustment allowed us to reduce the error rate. The detail of this technique is seen in Koike and Arai (2001). The percentage of cells with an error within 20 percent was increased to 70 percent of the total, and the percentage of cells with an error of more than 40 percent was significantly reduced.

5.3   Derivation of the Historical Population-Grid Data for Around 1890 in the Kanto Plain

The method used above for estimating HPD in East Biwa is rather simple, but it contains several time-consuming processes, for example, the adjustment for the arbitrary drawing of buildings, which is difficult to automate. Many basic GIS tools can reduce work time. We tried to develop and test several methods for HPD estimation using GIS tools. The manner of derivation of HPD for around 1890 (1890-HPD) in the Kanto Plain will be introduced in the following section.

5.3.1. Estimation Method

A basic resource in the HPD derivation was the *Jinsokuzu* map series. In these maps, where the concentration of buildings is low, each building is drawn separately. Where the concentration is high, adjacent buildings are drawn as a single, combined area. We represented concentrated areas as polygons. However, the number of separated buildings was too many to represent in this way, there being so many that the digitizing would have been very time consuming. To avoid this task, separated buildings were located as circles with an 8-meter radius centered upon them. This gave the average area occupied by a building and the resulting digitized data forms a vector-layer of buildings. Systematic sample points were generated with a 25-meter-wide grid over the vector-layer of buildings. Points placed in the concentrated areas of buildings were referred to as *Type A*, and those placed in the circles as *Type B*.

Population was estimated using a multiple-regression model. The dependent variable of the model is the population of each village P, which was obtained from a statistical report called *Chohatsu Bukken Ichiranhyo*. The independent variables are the number of *Type A* and *Type B* sample points (X_A, X_B). The following equation was employed:

$$P = aX_A + bX_B \tag{5.4}$$

where coefficients a and b are estimated from the data.

Using Equation (5.4), we obtained the HPD where the size of a spatial unit is a square 1 km^2, such as is used in the BGS system mentioned in Section 5.2.1. The population for a square cell was estimated using the number of sample points counted within it. The estimated 1890-HPD is shown in Figure 5.4. In this figure, the distribution of population is represented by the trend-surface technique.

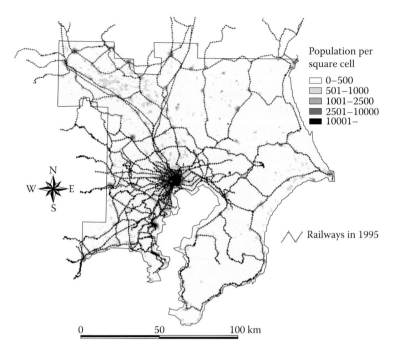

FIGURE 5.4
Trend surface based on the estimated grid population in 1890 (taken from Arai and Koike, 2003).

5.3.2   Estimation Accuracy

To examine the accuracy of the estimation, we compared the estimated village population obtained by Equation (5.4) with the actual population. Villages with an error rate under 30 percent accounted for 80 percent of the total. Although the accuracy was slightly lower than that obtained by the method described in the previous section, it was concluded that the above procedure is accurate enough for practical use.

5.4   Derivation of the Historical Population-Grid Data for Around 1930 in the Kanto Plain

This section shows a method of constructing the HPD for around the year 1930 (1930-HPD). Because the available maps and statistics are different, the method used is different from the above.

5.4.1 Source Materials

More accurate source topographical maps and population statistics are available for around the year 1930 when compared with the materials used in the preparation of the 1890-HPD. Standardized topographical maps at a scale of 1/50,000 covering the whole of Japan had already been completed and have been periodically updated. In addition, the *Kokusei Chosa* national population census had been started, and this provided accurate population data for the *Shi-Cho-Son* municipalities.

Despite the improved accuracy of source materials, there were two difficulties in the development of the HPD. First, the relationship between population and the area of buildings was more complicated than that addressed by the previous method, due to the change of population patterns and density of urban areas. Nonresidences or multistory buildings had increased in number following a trend away from traditional building practice toward Western-style construction and the modernization of land use. Second, the spatial units of population data were municipal districts that covered more than one village area. The methods used to generate the 1890-HPD could therefore not be used.

5.4.2 Estimation Method

In developing an alternative method for estimating the 1930-HPD, we divided all the municipal districts into three types: municipalities without an urbanized area, municipalities with a large urbanized area(s), and municipalities with a small urbanized area(s).

For municipalities without urbanized areas, an estimation equation could be created using the total area of buildings and the population. This method assumes that Equation (5.1) holds for small farming villages and employed the same procedure as in Section 5.3.

In the case of those municipalities with one or more large urbanized areas, we estimated the total population of the villages using Equation (5.1), and calculated the total population of the urbanized areas by subtracting the total population of the villages from the total population of the whole district.

When municipal districts had only small, urbanized parcels of less than 10 ha, we estimated the population density of those areas by applying a trend-surface analysis method developed in Koike (2002). This modification was employed because the result using the above method suggested that the error rates for municipal districts with small, urbanized areas tended to be very large.

Figure 5.5 shows the estimated population distribution around the year 1930.

5.4.3 Estimation Accuracy

Although the accuracy achieved throughout the whole estimation process could not be reviewed, it could be partially examined by comparing the

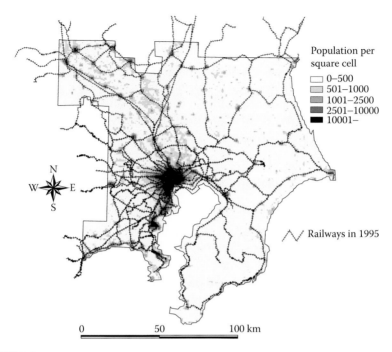

FIGURE 5.5
Trend surface based on the estimated grid population in 1930.

estimate with the actual municipal population. For the selected areas of the Saitama Prefecture and the Chiba Prefecture, 90 percent of municipalities were within an accuracy range of 30 percent.

5.5 Historical Population-Grid Database Covering the Period 1890–2000 in the Kanto Plain

By integrating the 1890-HPD and 1930-HPD with the existing population-grid data published by the Statistical Bureau of Japan, a HPD database was created covering the years 1890–2000. In this section, we outline this database and its applications.

5.5.1 Integration of the 1890-, 1930-, 1970-, and 2000-HPD

Since 1970, the Statistical Bureau of Japan has published population-grid data based on the National Population Census at five-year intervals. The spatial units of these grid data are square cells approximating 80 km, 10 km, and 1 km rectangles divided by latitude and longitude. The minimum data unit, a square 1 km², is a BGS, which was mentioned in Section 5.2. This system is widely used for various grid data other than population and was also employed for the HPD.

The HPD database was created by integrating the 1890-HPD, 1930-HPD, and the existing population-grid data for the years 1970 and 2000. This database works with ArcView and can be downloaded from the HPD Web site http://www.csis.v-tokyo.ac.jp/english/service/Prwas-PoP.html without charge for nonprofit uses.

5.5.2   Analysis of Population Change Using the 1890–2000-HPD

Two studies that use the 1890–2000-HPD are briefly mentioned below.

5.5.2.1   *Spatial Patterns of Population Change in the Kanto Plain*

This is a time–space analysis of historical population change. Figure 5.6 shows the spatial pattern of the population change from 1890 to 1930. A significant population increase is found in the area surrounding central Tokyo. The degree of increase is higher in the western part of the area than in the east. Although almost the whole area outside of central Tokyo maintained a steady population, several places experienced a degree of population decrease. Since population decreases were scattered over the whole study area, it is suggested that the central place system in the area was transformed from being traditionally road-oriented to being a modern railway-oriented one.

5.5.2.2   *Population Change Along Railway Lines*

The area of rapid population increase spread outward along railway lines, which suggests that from 1890 to 1930 the rail system played a significant role in Tokyo's urban growth. To examine this hypothesis, we overlaid the pattern of population distribution with railway networks.

Figure 5.7 shows the distribution of population in 1890 and in 1930 along the Takasaki Line, which stretches northwestward from central Tokyo. Although the Takasaki Line had not yet been completed in 1890, a main traditional road, called *Nakasendo*, ran alongside the Takasaki Line. A series of small population clusters lay along the road in 1890 at approximately 20-kilometer intervals, and these centers were traditional posttowns. In the distribution pattern for 1930, new settlements had developed between the established towns. The new population concentrations were around railway stations, and reflected the transformation of the central place system in the area.

The population distribution along the Chuo Line running westward from central Tokyo is shown in Figure 5.8. The Chuo Line was constructed in about 1890 through a sparsely populated area in the western part of the Kanto Plain. The scattered population pattern associated with the Takasaki

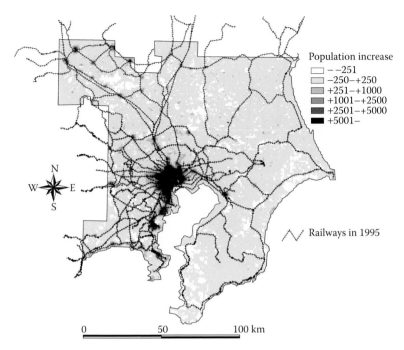

FIGURE 5.6
Spatial pattern of the population change during 1890–1930.

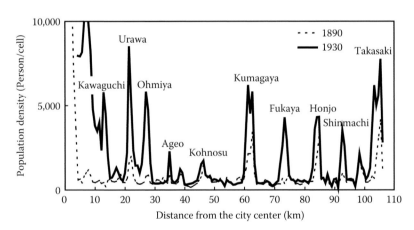

FIGURE 5.7
Population distribution along the Takasaki Line.

Line was not identified along the Chuo Line. A high density of development away from the city center to a point 15 km away took place until 1930. The area involved indicates the growth of suburban Tokyo.

The suburban area had expanded to a point 40 km away from the city center by 1970, and the highest concentration of people shifted from the city

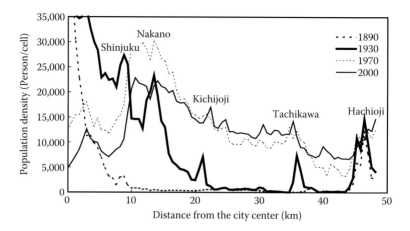

FIGURE 5.8
Population distribution along the Chuo Line.

center to 15 km away. The movement of the density peak corresponded with a noticeable population decline around the center. The location of the peak has not moved since 1970, although the height of the population peak has subsided.

5.6 Conclusion

We have developed a method for creating historical population-grid data in square 1 km² cells representing more than 100 years of change in the Kanto Plain. In principle, this method is applicable to any region in Japan and could be used in many other countries. In closing this chapter, we propose two applied studies to be undertaken with international collaboration.

The first proposal is to carry out a comparative study on the long-term urban growth of major world metropolises, focusing on detailed spatial-growth patterns using square 1 km² grid cells. If the population database for each metropolis is based upon the same grid, the effect of transportation infrastructure can be studied comparatively. Of particular interest is the difference in the urban growth characteristics of railway-oriented centers, such as shown in Section 5.5, when these are compared with the patterns seen in automobile-oriented cities.

A second proposed study is on rapid regional change in developing countries/regions. In the process of rapid economic growth, traditional farming villages have dramatically changed. A large-scale, statistical survey, such as a national population census, has the limitation that the data cannot keep up with the rapid social changes taking place during the survey interval. Remotely sensed data may be used to overcome this situation. In fact, as in

Section 5.2, we can estimate population from the area occupied by buildings observed in the remotely sensed images. Since remotely sensed data can be renewed at shorter intervals of time than a population census, they provide an effective means of monitoring the rapid regional changes in developing countries/regions.

In order to undertake the studies suggested above, international collaboration is indispensable. We anticipate that the 1890–2000-HPD database will be the first step toward such cooperative research.

References

Arai, Y. and Koike, S., Grid-based population distribution estimates from historical Japanese topographical maps using GIS, in *Modeling Geographical Systems: Statistical and Computational Applications*, Boots, B., Okabe, A., and Thomas, R., Eds., Kluwer, Dordrecht, 2003.

Koike, S., *Construction of a Historical Population Grid Database from Old Topographical Maps*, unpublished doctoral dissertation, the University of Tokyo, 2002 (in Japanese).

Koike, S. and Arai, Y., BGS population estimation from the topographical maps in the Meiji-Era, *Theor. Appl. GIS*, 9(1), 1–8, 2001 (in Japanese).

Lee, J., Redistributing the population: GIS adds value to historical demography, *Hist. Comput.* 8, 90–104, 1996.

Norton, W., Constructing abstract worlds of the past, *Geogr. Anal.*, 8, 269–288, 1976.

Siebert, L., Using GIS to document, visualize, and interpret Tokyo's spatial history, *Soc. Sci. Hist.*, 24, 537–574, 2000.

Taniuchi, T., Distribution of urban population in the Tokyo and Osaka metropolitan areas, 1883-1985, *The Proceedings of the Department of Humanities College of Arts and Sciences*, 101, 1995, pp. 99–118 (in Japanese).

6

Urban Employment Areas: Defining Japanese Metropolitan Areas and Constructing the Statistical Database for Them

Yoshitsugu Kanemoto and Reiji Kurima

CONTENTS

6.1 Introduction

For those interested in analyzing urban activity, the first task should be to define urban areas. The legal definition of a city is a natural starting point, but many urban activities extend beyond jurisdictional boundaries. For example, many workers in large metropolitan areas commute from suburban jurisdictions to central cities. We therefore need a definition of an urban area within which most everyday activities are undertaken. An urban area typically comprises a core area that has significant concentrations of employment, which is surrounded by densely settled areas that have close commuting ties to the core.

In the United States, the federal government has defined metropolitan areas since 1947 and provides a variety of statistical data relating to them. There is no counterpart in Japan, and the only definitions of metropolitan areas available are those proposed by a small number of researchers. Most of these adopt standards similar to the Standard Metropolitan Statistical Area (SMSA), which was in use in the 1960s and 1970s. However, in the U.S., two major changes have occurred since then, which reflect changes in the population distribution and activity patterns. First, in the 1980s, the Consolidated Metropolitan Statistical Area (CMSA) was introduced, which connects metropolitan areas that have significant interactions. Second, a new definition known as the Core-Based Statistical Area (CBSA) was introduced for the 2000 population census.

In Japan, changes in metropolitan areas motivated a revision of the first generation of metropolitan-area definitions. Kanemoto and Tokuoka (2002) proposed a new metropolitan-area definition to deal with complicated interaction patterns in Japanese metropolitan areas. The newly defined metropolitan areas are known as *Urban-Employment Areas* (UEAs), because they are based on employment patterns. The UEAs are divided between *Metropolitan-Employment Areas* (MEAs) and *Micropolitan-Employment Areas* (McEAs) according to their sizes. Researchers affiliated with the Center for Spatial Information Science at the University of Tokyo have been constructing a database for UEAs. In this chapter, we explain the definition of the UEA and a method of constructing an economic database for them.

6.2   The Need for a New Metropolitan-Area Definition

As noted above, a number of researchers have developed their own definitions of metropolitan areas. Examples are the Regional Economic Cluster (REC) of Glickman, the Functional Urban Core (FUC) of Kawashima, and the Standard Metropolitan Employment Area (SMEA) of Yamada and Tokuoka. These SMSA-type definitions apply different standards to central cities and suburban areas. According to the SMEA, a central city requires a population of at least 50,000, a percentage of nonagricultural workers of at least 75 percent, at least as many daytime occupants as nighttime ones, no more than 30 percent of the population commuting out, and no more than 15 percent commuting to another central city. A suburban municipality requires a percentage of nonagricultural workers of at least 75 percent and at least 10 percent of the population commuting to the central city.

The idea of defining central cities and suburban areas separately is attractive because of its simplicity. It first defines central cities, and then finds suburban areas for each of them, and the process does not involve iteration. However, it has shortcomings. For example, a city with a high population density may not be included in a metropolitan area. For example, Yamaguchi city, which is the capital of the Yamaguchi prefecture, did not belong to an SMEA until 1985. It was not classed as a central city because it had fewer daytime occupants than nighttime occupants and at the same time did not satisfy the conditions for being a suburb of another city.

Recently, this problem has become increasingly serious because of the emergence of a large number of subcenters and because of increasingly complex commuting patterns. If we use commuting ties to define a suburban area in relation to a particular central city, a city that is close to more than one central city may not belong to a metropolitan area. In the 1995 population census, there were 441 cities with populations of at least 50,000, of which 60 could not be classed as either central cities or suburbs of an SMEA. Of these 60 cities, 16 have populations of at least 100,000. Given that a single city with a population of 100,000 can itself be classed as an SMEA, excluding these cities from metropolitan areas is not consistent. Many cities that do not belong to an SMEA are suburban areas from which at least 30 percent of the population commutes out. Typically, commuters have more than one city to commute to. Almost 50 percent of these cities are located on the periphery of the Tokyo SMEA.

To deal with these problems, we can relax either the requirements for central cities or those for suburban areas. For example, the core of a metropolitan area may include subcenters with sufficiently large concentrations of employment even if they satisfy the requirements for classification as suburban areas of a central city. In the Tokyo metropolitan area, Yokohama, in which employment was about 1.4 million in 1995, could be included in the core area. Another possibility is to modify the requirements for suburban areas so that they take account of commuting to other suburban cities.

SMEAs have three types of requirement for central cities: namely, population size, urban characteristics, and the employment core. Of these elements, the employment-core requirements should be reexamined first so that areas with significant population concentrations are not excluded from metropolitan areas. Another problem with SMEA relates to the requirements for urban characteristics. The percentage of nonagricultural workers represents this element, but it is no longer an effective index of urbanization.

6.3 Metropolitan-Area Definitions in the U.S.

In revising the Japanese metropolitan-area definition, it is useful to study other countries that have experienced a similar trend of increasingly complex metropolitan areas. In the U.S., there was a major revision in 2000 with the introduction of a new metropolitan-area definition known the Core-Based Statistical Area (CBSA). According to the Office of Management and Budget (2000), a CBSA is a geographic entity associated with at least one core of 10,000 or more population, plus adjacent territory that has a high degree of social and economic integration with the core as measured by commuting ties. The standards designate and define two categories of CBSA: Metropolitan Statistical Areas and Micropolitan Statistical Areas (Office of Management and Budget, 2000, p. 82, 236).

A Metropolitan Statistical Area is associated with at least one urbanized area that has a population of at least 50,000, and a Micropolitan Statistical Area is associated with at least one urban cluster that has a population of at least 10,000 but less than 50,000.

A CBSA is identified in four steps. First, a CBSA must contain sufficiently large urban (densely settled) areas. Specifically, it must have an urbanized area, as defined by the Census Bureau, of at least 50,000 people, or an urban cluster, as defined by the Census Bureau, of at least 10,000 people.

Second, the core of a CBSA comprises a central county or counties associated with the urban areas. Specifically, a central county or counties must: (a) have at least 50 percent of its population in urban areas of at least 10,000 people; or (b) have within its boundaries a population of at least 5,000 located in a single urban area of at least 10,000 people.

Third, outlying counties of a CBSA must satisfy the commuting requirement that: (a) at least 25 percent of the employed residents of the county work in the central county or counties of the CBSA; or (b) at least 25 percent of the employment in the county is accounted for by workers who reside in the central county or counties of the CBSA.

Fourth, closely connected CBSAs are merged into one CBSA. In particular, two adjacent CBSAs merge into one CBSA if the central county or counties (as a group) of one CBSA qualify as outlying counties to the central county or counties (as a group) of the other CBSA using the measures and thresholds stated in (a) and (b) above.

Because of institutional differences, we cannot apply the U.S. definitions to Japanese cities. The most important difference is that the Japanese government does not define urban areas that extend beyond jurisdictional boundaries. The nearest equivalent in Japan is a Densely Inhabited District (DID) defined within a local municipality. The DID is defined by the Statistics Bureau as an area that is a group of contiguous Basic Unit Blocks, each of which has a population density of 4,000 inhabitants or more per square kilometer, or which has public, industrial, educational, and recreational facilities, and whose total population is 5,000 or more within a local municipality.

U.S. requirements for outlying areas have changed considerably. First, the measures of settlement structure, such as population density, that had been used to define outlying counties are no longer used; currently, commuting data are used. The reason for this change is that "as changes in settlement and commuting patterns as well as changes in communications technologies have occurred, settlement structure is no longer as reliable an indicator of metropolitan character as was previously the case" (Office of Management and Budget, 1999). In Japan, metropolitan areas have expanded into rural areas, and the use of settlement structure may no longer be relevant. Second, the percentage commuting out was raised from 15 percent to 25 percent, because the percentage of workers in the U.S. who commute to work outside their counties of residence increased from approximately 15 percent in 1960 to almost 25 percent in 1990. In Japan, we should also reconsider the commuting ratio. However, it is not clear whether the ratio should be raised, because, as we explain later, commuting patterns in Japan are much more complicated than in the U.S.

6.4   The Structure of Japanese Metropolitan Areas

According to the 1995 population census, 724 cities and towns have DID populations of at least 10,000, of which 440 have total populations of at least 50,000. The number of cities with DID populations of at least 50,000 is 297.

Many urban areas are dormitory towns, and relatively few of these are employment centers. Of the 724 (297) cities and towns with DID populations of at least 10,000 (50,000), only 281 (120) have larger commuter inflows than outflows. Some large cities, such as Yokohama, Chiba, and Kawasaki, have larger outflows than inflows. However, the central wards of these cities are employment centers that have larger commuter inflows than outflows and DID populations of at least 50,000.

Consider commuting patterns. The average commuter-outflow proportion is 32 percent, but employed residents commute to a wide variety of urban areas. Some cities have highly concentrated commuting patterns: More than 50 percent of the employed residents of 16 cities and towns (Tomiya, Wako, Urayasu, Fuchu, Komae, Hoya, Nagayo, Kokufu, Uchinada, Ichikawa, Sanwa, Kouyagi, Ishikari, Musashino, Hashikami, and Yakumo) commute to one city. Another extreme is Zama, where more than 5 percent of the employed residents commute to one of six other municipalities.

6.5   Defining Urban-Employment Areas

Kanemoto and Tokuoka (2002) proposed a new metropolitan-area definition known as the Urban Employment Area (UEA). UEAs are divided between Metropolitan Employment Areas (MEAs) and Micropolitan Employment Areas (McEAs) according to their sizes. These are similar to the CBSA for U.S. cities, but there are substantial differences in specific requirements to reflect the higher densities and more complicated commuting patterns in Japanese cities.

In defining metropolitan areas for Japan, we must take the following four conditions as given.

1. The building blocks of metropolitan areas are municipalities (cities, towns, and villages), because most statistical data are only available up to the municipality level.

2. The prevalence of commuting by car in smaller metropolitan areas has increased the numbers commuting to cities from areas of low population density. The use of population-density standards in defining outlying areas is no longer practical.

3. In identifying densely inhabited urban areas, we use DID populations.

4. The inclusion of outlying suburban areas is determined by commuting patterns between municipalities.

After examining many alternatives, Kanemoto and Tokuoka (2002) proposed the following definition of a UEA. First, local municipalities (cities, towns, and villages) are the building blocks of UEAs. The core of a UEA is a collection of densely settled municipalities (i.e., those with DID populations of at least 10,000) that do not constitute the "outlying municipalities" (suburbs) of any other core. The outlying municipalities of a UEA are defined mainly by the requirement that at least 10 percent of employed workers commute to the core. An MEA is a UEA whose core has a DID population of at least 50,000. An McEA is a UEA whose core has a DID population of at least 10,000 and less than 50,000. More specifically, they use the following standards.

6.5.1 Requirements for a Core

1. In the first round, potential cores are municipalities with DID populations of at least 10,000.

2. A municipality that is an outlying area of another central city is excluded from being a core.

3. For a pair of municipalities that each satisfy the requirement for being in a core and also satisfy the commuting-ratio requirement for being an outlying area of the other municipality, the one with the lowest commuting ratio is in the core, and the other is its outlying area.

4. An outlying municipality is included in the core if the following two requirements are satisfied. A "major" city (known as a *Seirei Shitei Toshi*) for which data on its wards are available is included in the core if at least one of its wards satisfies the following requirements.

a. The employees-to-residents ratio (i.e., the ratio of the number of employees to the number of residents) is at least unity.
b. The DID population is at least 100,000 or one-third of the core.

Condition 4 implies that a core may contain more than one municipality. The reason for adding the condition on the size of the DID in 4b is that the largest metropolitan areas, such as Tokyo and Osaka, have multiple central cities, and some very small municipalities satisfy the employees-to-residents ratio. It is inappropriate to include a municipality of 5,000 in the core alongside the central city of Tokyo, which has more than 7 million employees.

6.5.2   Requirements for an Outlying Area

1. A municipality is an outlying area of a core if at least 10 percent of its employed residents work in the core.
2. If a municipality satisfies condition 1 for more than one core, it is included in the outlying area of the core with which it has the strongest commuting ties.
3. A second-order outlying municipality that is an outlying area of another outlying municipality is included in a UEA. Higher-order outlying municipalities (i.e., third-order, fourth-order, etc.) are also included in UEAs. The criterion for a second-order municipality is that, of all the target municipalities, its commuting ratio to a first-order outlying municipality is the highest and satisfies the 10 percent criterion. Higher-order outlying municipalities are defined analogously.
4. If a municipality simultaneously satisfies the requirement for being an outlying area of a core and the requirement for being another outlying municipality, it is classified as an outlying area of the one with which it has the highest commuting ratio. That is, if 16 percent of the employed residents in city A work in core B and 17 percent of them work in city C, which is an outlying municipality of core B, then city A is an outlying area of city C.

6.5.3   The Iterative Procedure for Defining UEAs

UEAs are determined by the following iterative procedure.

6.5.3.1   The First Iteration

1. Choose municipalities with DID populations of at least 10,000 as potential central cities.

2. Exclude as potential central cities defined by 1, those municipalities that are outlying areas of other potential central cities.

3. Determine the outlying municipalities for the potential central cities by using the following procedure.

   a. Select municipalities for which the percentage of employed residents who work in a central municipality is at least 10 percent; and for each of them, determine the central municipality that has the highest commuting ratio. This identifies potential first-order outlying municipalities.

   b. Determine potential second-order outlying municipalities by choosing municipalities that satisfy the commuting-ratio criterion.

   c. Determine potential second-order outlying municipalities.

   d. Determine potential third-order outlying municipalities.

   e. Check for fourth-order outlying municipalities. (Currently, there are none.)

   f. If a municipality is simultaneously a first-order, second-order, or third-order outlying area, identify the target municipality with the highest commuting ratio.

   g. List the outlying municipalities for each central city.

6.5.3.2   The Second Iteration

1. Of the potential outlying municipalities identified in the first iteration, those that satisfy the following two requirements are included in the cores of the UEAs to which they belong. If a candidate is a "major" city, it is included in the core if at least one of its wards satisfies the following requirements.

   a. The employees-to-residents ratio is at least unity.

   b. The DID population is at least 100,000 or at least one-third of that of the central municipality.

2. Potential first-order outlying municipalities are those in which at least 10 percent of employed residents work in the central municipality. For each of these, choose the central municipality that has the highest commuting ratio. The procedures applied in the first iteration are applied to determine higher-order outlying municipalities.

6.5.3.3   Other Iterations

The procedures used in the second iteration are used for subsequent iterations. Step 1b in the second iteration for adding an outlying municipality to the core remains the same. In particular, the central municipality is the one

FIGURE 6.1
The Metropolitan Employment Areas for the 1995 population census.

identified in the first iteration and does not include those added in the second iteration.

6.6   Urban-Employment Areas for the 1995 Population

For the 1995 population census, three iterations were needed to define the UEAs. The total number of UEAs is 278, of which 118 are MEAs and 160 are McEAs. Figure 6.1 maps Japan's MEAs. Table 6.1 presents a rough outline of the requirements of UEAs. Figure 6.2 shows MEAs, McEAs, DIDs, and municipality boundaries in Ibaraki Prefecture. The darkest gray areas within MEAs and McEAs are DIDs. The second-darkest areas are the cores of MEAs and McEAs, and light-gray areas are their outlying areas. Light-gray curves show boundaries of municipalities (cities, towns, and villages).

TABLE 6.1

Requirements of Urban Employment Areas

Requirement	Urban Employment Area
Categories	Metropolitan Employment Area: The DID population of the core is at least 50,000. Micropolitan Employment Area: The DID population of the core is at least 10,000 and less than 50,000.
Qualification of Areas	City of at least 10,000 DID people.
Qualification of Central Municipalities (Cores)	Municipalities that satisfy either of the following two requirements are included in the core. (The core may include more than one municipality.) (a) The DID population is at least 10,000, and the municipality is not an outlying area of another core. (b) The requirements for an outlying area are satisfied, and the following two requirements are also satisfied. 　(i) The employees-to-residents ratio is at least unity. 　(ii) The DID population is at least 100,000 or one-third of the core. For a pair of municipalities, each of which satisfies the requirement for being in a core and also satisfies the commuting-ratio requirement for being an outlying area of the other municipality, the one with the lowest commuting ratio is in the core, and the other is its outlying area.
Qualification of Outlying Municipalities	(a) A municipality is a first-order outlying area of a core if at least 10 percent of its employed residents work in the core. (b) A municipality is a second-order outlying area of a core if at least 10 percent of its employed residents work in an outlying municipality and the commuting ratio to the municipality is the highest among all target municipalities. Higher-order (i.e., third-order, fourth-order, etc.) outlying municipalities are defined analogously. (c) If a municipality satisfies requirement (a) for more than one core, it is included in the outlying area of the core with which it has the strongest commuting ties. (d) If a municipality satisfies the requirement for an outlying area of a core, as well as for another outlying municipality, it is an outlying area of the one with which it has the highest commuting ratio.

6.7 The Construction of the MEA Economic Database

Since the building blocks of the UEAs are local municipalities, the UEA data can be obtained by summing the municipality-level data. However, data on many important economic indicators are not available at the municipality level. For example, most production data are available only at the prefectural level. Kurima and Ohkawara (2001) constructed MEA data for total production (value added), private capital, and social-overhead capital.

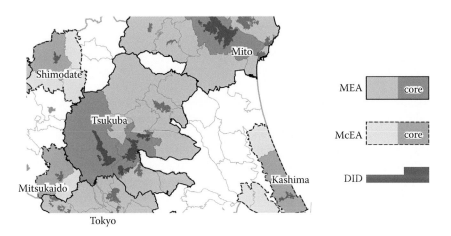

FIGURE 6.2
The relationship among MEAs, McEAs, DIDs, and municipalities: An example of Ibaraki Prefecture.

The Annual Report on the Prefectural Accounts includes detailed production data for 47 prefectures and major cities (11 cities up to 1988 and 12 from 1989). Private capital (for manufacturing and nonmanufacturing industries) and social-overhead capital (for 12 industries) are available for prefectures (but not for major cities). Kurima and Ohkawara (2001) construct MEA data by combining these data with the employment data for local municipalities.

First, note that an MEA is a collection of local municipalities that may belong to different prefectures. Kurima and Ohkawara (2001) allocate the prefecture-level data to municipalities by using proportional allotment and then aggregate them to obtain MEA data. In prefectures that include the major cities, they construct the data for the remaining areas and allocate them to municipalities outside those cities. The production data are available for 14 industrial categories (12 industries, the public sector, and others). The output of each industry is allocated to municipalities on the basis of the industry employment shares. For example, consider an MEA denoted by A. Denote the number of workers in industry i in municipality j by $N(i,j)$, and denote the total production and employment of the prefecture (denoted by I) that contains the MEA by $Y(I,i)$ and $N(I,i)$, respectively. Total production in the MEA is then

$$Y(A) = \sum_j \sum_i Y(I,i) \frac{N(i,j)}{N(I,j)}$$

Second, the private capital in each MEA is obtained by using proportional allotment on the basis of the production shares in manufacturing and nonmanufacturing industries (rather than employment shares).

Third, different allotment procedures are used for each of the four types of social-overhead capital. In agriculture, forestry, and fishing, social-overhead capital is allocated on the basis of production shares in the agricultural sector. In industrial infrastructure, allocation is based on the production shares in the manufacturing industry. In telecommunications and railways, allocation is based on total production. Allocation of infrastructure for residents, such as parks and neighborhood streets, is based on population.

We have annual data on production and the capital stock, but census population and employment data are collected only every five years. Annual population data are available from the Basic Resident Registers, but their definitions differ slightly from those of the more detailed census data. The annual data are constructed by using the Resident Registers data to modify the simple linear interpolations of the census data. Let P denote the census population data and let Q denote the Resident Registers data. The annual data that we use, $\hat{P}_{n+i}$, are then

$$\hat{P}_{n+i} = P_n + \frac{i}{5}(P_{n+5} - P_n)\frac{Q_{n+i}}{\hat{Q}_{n+i}}$$

where n is the census year, $i = 1,2,3,4$, and

$$\hat{Q}_{n+i} = Q_n + \frac{i}{5}(Q_{n+5} - Q_n)$$

The Annual Report on Prefecture Accounts contains annual employment data for three industry categories — namely, primary, secondary, and tertiary — but these are only available for prefectures and 12 large cities. We use these data for Q in the procedure described above.

The original data sources for the MEA economic database are as follows.

Number of employees and population: Population Censuses of 1980, 1985, 1990, and 1995, and Basic Resident Registers (Jumin Kihon Daicho) 1980–1995.

Production (Value Added): Annual Report on the Prefecture Accounts.

Private Capital Stock and Social Overhead Capital: Estimates by the Central Research Institute of the Electric Power Industry (CRIEPI). (The method of estimation is explained in Ohkawara et al., 1985.)

6.8   Conclusion

We have developed an urban-area definition called the UEA for Japanese cities. The UEAs are divided between MEAs (large UEAs) and McEAs (small UEAs) according to their sizes. We have also constructed an economic database for the MEAs. The UEAs for the population censuses of 1980, 1990, 1995, and 2000 and the MEA economic database for the 1995 definition can be found on the UEA Web site (www.urban.e.u-tokyo.ac.jp/UEA/index_e.htm). A number of researchers and government agencies have already used the UEA. Chapter 8 of this book contains an example of empirical studies that use the MEA economic database.

Acknowledgment

This research was supported by Grant-in-Aid for Scientific Research No. 10202202 and no.1661002 from the Ministry of Education, Culture, Sports, Science, and Technology.

References

Kanemoto, Y. and K. Tokuoka, Proposal for the Standards of Metropolitan Areas of Japan, *J. Appl. Region. Sci.,* 7, 1–15, 2002 (in Japanese).

Kurima, R. and T. Ohkawara, Construction of MEA-based Economic Data, mimeo, 2001 (in Japanese).

Office of Management and Budget, Recommendations From the Metropolitan Area Standards Review Committee to the Office of Management and Budget Concerning Changes to the Standards for Defining Metropolitan Areas, *Federal Register,* 64(202), October 20, 1999.

Office of Management and Budget, Standards for Defining Metropolitan and Micropolitan Statistical Areas, *Federal Register,* 65(249) December 27, 2000.

Ohkawara, T., Matsu-ura, Y., and Chuma, M., Chiiki Keizai Data no Kaihatsu Sono 1 — Seizougyou Shihon Sutokku to Shakai Shihon Sutokku no Suikei (Estimation of Regional Economic Data Part 1: Manufacturing Capital and Social Overhead Capital), Central Research Institute of Electric Power Industry, Report No. 585005, 1985 (in Japanese).

7

Data Modeling of Archaeological Sites Using a Unified Modeling Language

Teruko Usui, Susumu Morimoto, Yoshiyuki Murao and Keiji Shimizu

CONTENTS

7.1 Introduction

This chapter illustrates a data model for archaeological sites that enables exchange of data among archaeological communities around the world. The first section describes the nature of archaeological site data. The second section shows the difference between the Japanese data model and the Western data model (i.e., the Harris Matrix model). The third section discusses an object-oriented model for recording archaeological site data in comparison with the traditional layer-based model. This section also explains the procedure for this modeling and a method of implementing it with the Unified Modeling Language (UML). The fourth section applies the UML to both the Japanese data model and the Harris Matrix model. The sixth section concludes the chapter with remarks on the common data model that can be shared with researchers throughout the world.

7.2   Characteristics of Archaeological Information and a Site Survey

Archaeological sites represent evidence of human activities in the past. This evidence can be classified roughly into two categories: namely, *archaeological features* and *artifacts*. Postholes and moats are examples of *archaeological features*, which exist in a certain location or as a part of the ground, and which are basically not transferable. Stone tools and earthenware come into the category of *artifacts,* which are transferable. The place in which artifacts and remains are excavated is called an *archaeological site*. For archaeologists, it is the collected information provided by artifacts and remains at archaeological sites that is the most essential resource to investigate human activities in the past.

In archaeology, there are various kinds of surveys, such as distribution surveys, site surveys, trench surveys, and excavation, and the results of those surveys are finalized in reports. During excavation, it is important to record precise positional relationships, configuration and position of remains, and location and direction of artifacts. The drawing of archaeological features, as shown in Figure 7.1, provides spatial information and positional relationship of remains and artifacts in a survey report.

Thus, Geographic Information Systems (GIS) play a significant role in the management and analysis of archaeological information that contains geographical information (Wheatley and Gillings, 2002).

However, there is no standardized procedure by which information is collected, as collection procedures depend on the decisions made by the excavating archaeologists. Whether to interpret an excavated hole as a pillar hole or not is dependent on the knowledge of excavation teams. Furthermore, after excavation, the sites are most commonly covered with soil or building constructions, and the information becomes available only in a report, with drawings of archaeological features and photos taken. Information sharing requires the establishment of standardized recording methods and a database structure reflecting the least subjective interpretation. Standardization is required because of the differences in the approach taken by

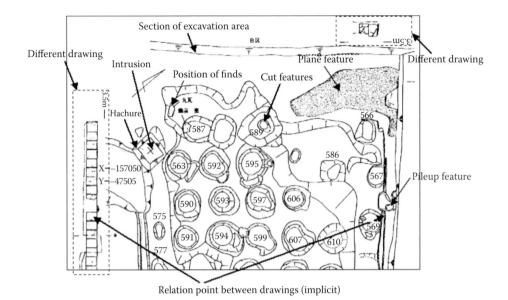

FIGURE 7.1
Drawing of archaeological features.

archaeologists in Japan and Europe in the preservation and recording of archaeological information.

7.3 Differences between Japanese and European Techniques in Data Recording and Organizing Archaeological Survey Data

Survey systems and data-recording techniques are significantly different in Japan and Europe. In Europe, the differences of stratification are classified into units of stratification based upon stratigraphy, and each unit of stratification is precisely surveyed with repeated observations of stratigraphic sequences. Then, the remains are objectively reported in a stratigraphic sequence diagram, generally called a Harris Matrix (Harris, 1989).

Figure 7.2 shows a Harris Matrix diagram. From the aspect of information recording, it has superiority in the adoption of the minimum unit based on types of soil, which is least influenced by arbitrary decisions of excavation teams. The numbering 115 to 153 in Figure 7.2 indicates the relationship of stratigraphic sequences during excavation. The recording method enables archaeologists to reproduce excavation processes with possible interpretations. In contrast, repeat processes are unobtainable after excavation by the Japanese recording methods shown in Figure 7.1.

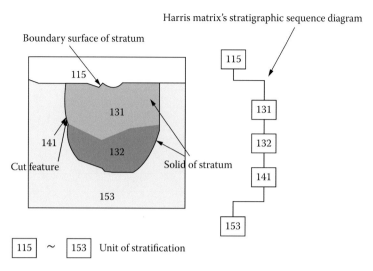

FIGURE 7.2
Harris Matrix's stratigraphic structure and sequence diagram.

On the other hand, Japanese archaeologists first identify a feature surface, which becomes the basis of the survey, and each piece of the remains is examined based upon geological transitions relative to the feature surface. The result is reported in the drawing of archaeological features. Compared to the European stratigraphic technique, objective reporting on remains in the upper layers is basically left out of the Japanese surveys, because information recording is determined on site. In Japan, extensive surveys mostly take place in a relatively hot and humid environment, and such techniques enable archaeologists to retain efficiency of surveys and maintain quality. The boundary of stratification has significant meaning in archaeology, and its two-dimensional diagram is considered a plainer representation of remains. The clarification of the relationship between stratigraphic sequence diagrams and drawings of archaeological features enables database development and integration of archaeological information collected in both Japan and Europe. Consequently, archaeological information sharing could become feasible, allowing for the shared use of archaeological information to proceed worldwide.

For that purpose, we propose that it is critical to articulate the relationship between the Harris Matrix stratigraphic-sequence diagram and the Japanese drawing of archaeological features, and to define a schema for an archaeological-information database to identify the context and structure of archaeological information. However, the layer structure in the existing GIS model has no flexibility to fully incorporate association and definition of archaeological information. Given that fact, we consider that instead of the layer-based model, it is beneficial to adapt the feature-based GIS data model to object-oriented GIS technology — a rapidly advancing technology.

7.4 Object-Oriented GIS and an Archaeological-Information Database

7.4.1 Two Kinds of GIS Data Models

Existing GIS has a database structure derived from paper maps, which require overlaying of several outlines, showing such features as buildings, roads, and administrative boundaries. In a similar way, GIS adopts the same layer structure, and geographic spaces are represented with the overlay technique. As shown in Figure 7.3, general database structure supports layers consisting of geometric and attribute databases, ensuring the collated data is merged and combined in a spatial index. The layer-based data model has an interlayering relationship problem, which can be significant. For instance, in the electricity-management system, electric line (line), power pole (point), and power plant (polygon) layers are created and manipulated in electricity flow and facilities. In this layer-based data model, realistic situations often occur. For example, the electricity line remains even if a particular pole in the layer is erased. Since the mid-1980s, a more robust, feature-based data model has been operational, superseding the layer-based data model (Tang et al., 1996). This development has been accelerated by the object-oriented, technological advance leading to the standardization of geographical information by the International Organization for Standardization (ISO) Technical Committee (TC 211, Geographic

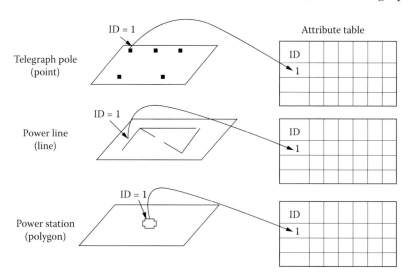

FIGURE 7.3
The structure of a layer-based data model.

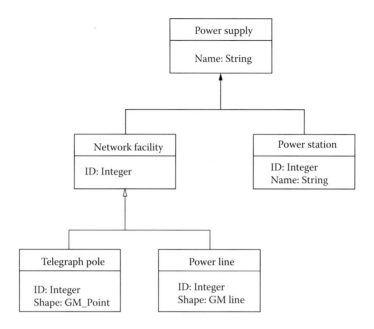

FIGURE 7.4
The application schema of a feature-based data model.

information/Geomatics). The feature-based data model has been influenced by the object-oriented GIS technology.

Figure 7.4 shows the differences between feature-based and layer-based models in the process of defining database structure or schema. The layer-based model generates layers consisting of geometric database and attribute database. On the other hand, the development of the database structure or schema of the feature-based model involves defining the feature type followed by the relationships between each type. For example, in the electricity-management system, the features of power pole, electric line, and power plant are identified, followed by the relationships between the features. Eventually, a database schema is defined by itself with the definitions. It is the geographic information standards that set such feature definitions and the rules of relationships between features. This is the first step in defining archaeological features based on geographic-information standards to develop a database structure of archaeological information.

7.4.2   Standardization of Geographic Information and UML

The purpose of standardizing geographic information is the implementation of information sharing and its interoperability. In the area of object-based GIS, standardization does not simply imply integration of data formats. Specifically, it defines the attributes, operations, and associations of physical features, such as roads, buildings, and archaeological sites. Their semantic attributes and operations are encapsulated into feature classes, such as roads, and the implementation of common rules that enable the sharing and mutual utilization of the information. The technique of defining feature classes and class relationships in geographic information is called *data modeling*. In the standardization of geographic information, common rules for data modeling are specified with a special language called *Unified Modeling Language* (UML).

Following ISO/TC211, UML, a language for object-based technique, is recognized as the conceptual schema language for standardizing geographic information. UML was originally developed by Grady Booch, Ivar Jacobson, and James Rumbaugh of the Rational Software Corp. in the United States and introduced as *Object-Modeling Technique* (OMT), a technique that uses diagram representations. Version 1.1 was certified as a standard language of the Object Management Group (OMG) in November 1997. Unlike other object-oriented languages, such as C++ and Java, the UML is a visual-modeling language depicting a diagram to define objects and identify any relationships among them. At the same time, it enables the creation of a metamodel integrating notations and semantics (Worboys, 1994). This chapter introduces the research findings in the data modeling of archaeological information with the aim of effective information sharing and utilization in the field of archaeology. The modeling was conducted based upon the geographic-information standards defined by ISO/TC211.

The organizational head office of ISO/TC211 (www.isotc211/) is currently located in Norway, and its Japanese contact for the standardization of geographic information is at the Geographical Survey Institute (GSI). In 1999, the GSI produced the Japanese Standards for Geographical Information 1.0 (JSGI 1.0), which was the result of a public–private, collaborative research partnership that began in 1996. In 2002, the GSI released the second version of the JSGI on the Internet.

7.4.3   Data Modeling of Archaeological Information and the General-Feature Model

Geographic-information standards have a characteristic in defining the structure of the GIS database with a conceptual model generating real-world abstraction. This conceptual model is the *General-Feature Model* (GFM). Figure 7.5 provides a clear picture of the Domain Reference Model, consisting of four levels, including the GFM. Ancient remains are classified into features, and a *Feature Catalogue*, called the "Feature Dictionary," is created to clearly define the features. An application schema is developed using the UML for digitization of the features. A diagram is represented in UML as a schema, which provides the framework and content of archaeological information to be stored in a computer.

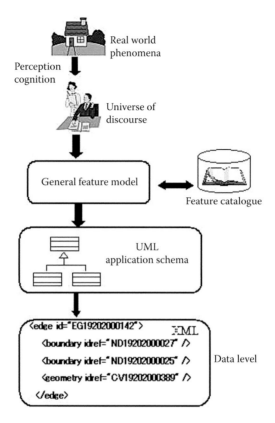

FIGURE 7.5
Domain reference model in a general feature model.

In cognitive linguistics, *discourse* means language communication. In fact, the world in which human beings are engaged in language communication is considered the universe of discourse. In short, the universe of discourse means the real world in which entities and phenomena are understood and explained by language. The objects derived from the processes of abstraction and classification of entities and phenomena are called *features*. The world in which we communicate about ancient remains represents the universe of discourse on ancient remains. Such communication is established by use of universal meanings; in this case, technical terms in archaeology. We human beings understand ancient remains in dictionary form, and for information sharing in GIS, it is essential to generate the feature catalogue of archaeological information and have the meanings and structure understood through the dictionary. The key point is that the dictionary should be usable on a computer. To that end, archaeological features are defined by the UML so that an application schema, the structure of the database, is consequently determined (Peckham and Lloyd, 2003).

The Domain Reference Model (DRM), shown in Figure 7.5, indicates the basis of archaeological data modeling. The DRM consists of four distinct levels.

1. The first level is the conceptual model, which extracts the universe of discourse on ancient remains from the real world.
2. The second level is the GFM, which abstracts the archaeological features and then creates a catalogue of archaeological information.
3. The third level is the application schema, which depicts the content and framework of archaeological information using the UML.
4. Finally, the data level implements geometric and topological spatial objects as specific spatial datasets. In this level, the data are encoded using XML (Usui, 2003).

In Japanese archaeological surveys, once a survey is completed, the remains are returned to their original state. Meanwhile, excavated artifacts are removed and kept in a separate place. Since the information on positional relationships of artifacts and remains are lost after the survey, a report becomes invaluable as the only information for archaeologists. Moreover, given that the survey involves excavating multiple soil layers, the remains in the upper soil layers need to be removed to reach those in the lower layers. Thus, the downward excavating process suggests that the remains found in the upper soil layers could not be restored to their original form. For this reason, a survey report and drawing of archaeological features must contain all the necessary information, especially the drawing of archaeological information, which would be required to define the archaeological features.

The geographic-information standards of ISO 19109, Rules for Application Schema, specify the way to define objects and the spatial relationship between features. This gives the impression that the standards provide specific methods for defining objects, but this is not so. In fact, the ISO 19109 Rules for Application Schema employs the UML to define objects, thus enabling the integration of general-information systems and GIS. Moreover, with the application of geographic-information standards, the defining processes of archaeological feature shapes and time attributes become simple. Both the Spatial Schema — defined by ISO 19107 — and the Temporal Schema — defined by ISO 19108 — form shape and time components or classes in the model, respectively.

Time is a critical element in archaeological information. The data may give a clue to a specific calendar year or a certain era; or, in some cases, no identifiable information at all. By applying these standards to archaeological information, it became feasible to make use of time-defining methods in addition to spatial information. Table 7.1 introduces object data types defined in ISO 19107 Spatial Schema and ISO 19108 Temporal Schema.

TABLE 7.1

Major Data Types of the Geographic Information Standards

Data Type	Representation
GM_Point	Spatial location (point)
GM_Curve	Spatial curve line
GM_Surface	Spatial curved surface
TM_Instant	Temporal position (time)
TM_Period	Temporal line (period)
Character String	Nearly identical to string type and character set addressable

7.5 European Stratigraphic-Sequence Diagrams Using the Harris Matrix and UML Modeling on Japanese Drawings of Archaeological Features

7.5.1 Class Representing the Archaeological Site (*Archaeological Site Class*)

The core pieces of information from archaeological sites are archaeological features and finds that are collected during a survey and recorded in a survey report. The report contains drawings of archaeological features, and maps indicating the most critical findings, such as shape, location, direction, and relative position of archaeological features. All the spatial attributes are provided in the drawings of archaeological features. Thus, in Japan, to compile a database, modeling becomes critical to accomplishing information sharing.

Figure 7.6 shows the definitions of an archaeological site class in a UML diagram. The class representing archaeological sites is the most significant in explaining the whole archaeological site. There are seven archaeological class attributes: identification number (identifierOfSite), name (nameOfSite), address (addressOfSite), duration (periodOfSite), area (archaeologicalArea), descriptions, and other information (additionalAttribute). In addition, there is a site-owner class (LandOwner), administrator class (AdministratorOfSite), survey-finding class (ResultOfInvestigation), and structure class (StratigraphicStructure). The archaeological site class and those four classes are parts of the whole. The relationship between these classes is considered "composition," since the components are all deleted in the case of taking out the whole archaeological site. In the figure, the class relations are drawn in filled rhombus.

In Japanese archaeological surveys, research findings (ResultOfInvestigation class) are completed with drawings of archaeological features; on the other hand, in the case of overseas surveys, stratigraphic-sequence diagrams

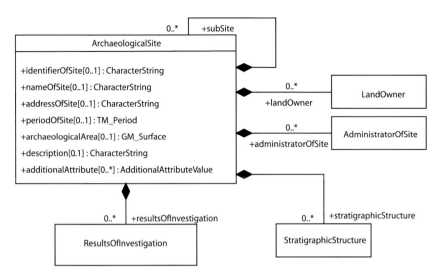

FIGURE 7.6
UML diagram of an archaeological site.

are produced. The diagrams depict the lowest-level configuration of stratigraphic structure (StratigraphicStructure class). The data types of the geographic-information standards, shown in Table 7.1, are applied to the archaeological-site properties. The data type of duration (periodOfSite) is the same as TM_Period. The duration of archaeological sites can be identified in a calendar year in some cases; in other cases, only its period can be estimated. The data types enable us to suggest the Jurassic and Cretaceous periods, which provide uncertain time periods, with no specific beginning and end years, in addition to the Gregorian and Japanese calendars. The archaeological sites certainly have addresses (addressOfSite), and basic information is recorded in the section. To provide precise location information of archaeological sites, GM_Surface, a data type of the geographic-information standards, is used.

7.5.2 Drawing of Archaeological Features and Stratigraphic-Sequence Diagram

Figure 7.7 explains the relationship between the drawings of archaeological features developed in Japanese archaeological surveys and the stratigraphic-sequence diagram, or Harris Matrix, of the European surveys. The archaeological-site class is made of aggregation of multiple soil stratifications. As a result, an attribute StratigraphicStructure class is used to describe the stratigraphical structures. This class also has components of SolidOfStratum class, indicating configuration of each stratigraphy, and BoundarySurfaceOfStratum, expressing the boundary surface of the stratum. UnitOfStratification class, a component of the Harris Matrix stratigraphic sequence diagram class,

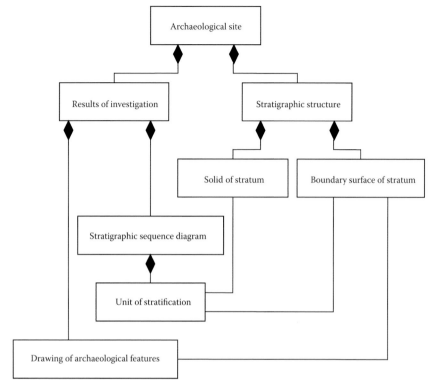

FIGURE 7.7
Relationships between Japanese drawings of archaeological features and the European Harris Matrix stratigraphic Sequence diagram.

has association with SolidOfStratum class and BoundarySurfaceOfStratum class. On the contrary, the drawing of archaeological features is a projection of the boundary surface of the stratum. The class of BoundarySurfaceOfStratum can define the relationship between two systems: the drawing of archaeological features and the stratigraphic-sequence diagram.

The drawing of archaeological features plays a critical role in archaeological surveys. The representation of the drawings is the DrawingOfArchaeologicalFeatures class, which is defined in Figure 7.8. The class represents the entire survey findings, which consist of ground plan, cross-section, and side view. As introduced in Figure 7.1, the contents of the drawing of archaeological features are categorized into the figures of archaeological feature, excavation area, intrusion, declaration, and reference point, and they have the relationship of aggregation. The most significant element in the drawing of archaeological features is an ArchaeologicalFeature class. This class enables us to classify whether it has a dent, swell, or plane surface relative to the boundary surface of the stratum. Each condition is named as an archaeological cut feature, archaeological pile-up feature, and archaeological plane feature.

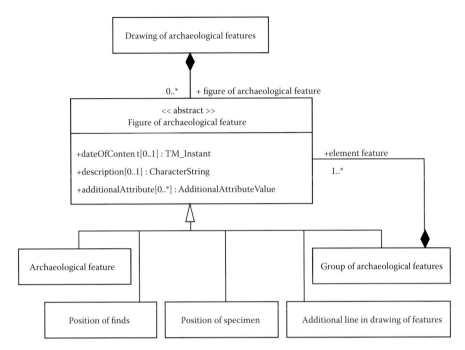

FIGURE 7.8
UML diagram of Archaeological Feature class.

7.6 Conclusion

For the worldwide implementation of archaeological information sharing, it is necessary to compare various information contexts and structures by countries and organize them based on specific rules. The object-oriented GIS have the rules of the geographic-information standards. Features can be defined by the UML so that the definitions of database schema become attainable. Unlike existing GIS, the object-oriented GIS provides the feature objects in geographic information, which are the smallest units consisting of geographic spaces, and treat them as if they are components of engineering products. The geographic-information standards can be considered as rules defining feature standards. This chapter has described the database-production process of archaeological information that meets the geographic-information standards. The features consisting of archaeological information are classified into two major classes: ArchaeologicalSite class, which is at the upper level; and ArchaeologicalFeature class, which can be subclassified into archaeological cut, archaeological pile-up, archaeological plane, and soil-layer classes, as shown in Figure 7.1. The ArchaeologicalFeature class corresponds to the boundary surface of the stratum in the Harris Matrix, and it enables the management and searching of a single database containing archaeological information collected in Japan, Britain, and the U.S.A. The object-oriented GIS assume features as the component units in geographic spaces and achieve data sharing by creating a feature catalogue based on the rules. In fact, the recognition process adapted to GIS in this chapter is similar to that used by human beings to understand the meanings of phenomena through dictionaries.

References

Wheatley, D. and Gillings, M., *Spatial Technology and Archaeology: The Archaeological Applications of GIS*, Taylor & Francis, London, 2002.

Harris, E.C., *Principles of Archaeological Stratigraphy*, 2nd ed., Academic Press, London, 1989.

Tang, A.Y., Adams, T.M., and Usery, E.L., A spatial data model design for feature-based geographical information systems, *Int. J. Geogr. Info. Syst.*, 10(5), 643–659, 1996.

Worboys, M.F., Object-oriented approaches to geo-referenced information. *Int. J. Geogr. Info. Syst*, 8(4), 385–399, 1994.

Peckham, J. and Lloyd, S.J., Eds., *Practicing Software Engineering in the 21st Century*, IRM Press, Hershey, 2003.

Usui, T., GIS revolution and geography — object oriented GIS and the methodology of chorography, *Geogr. Rev. Jpn.*, 76(10), 687–702, 2003.

8

How to Find Free Software Packages for Spatial Analysis via the Internet

Atsuyuki Okabe, Atsushi Masuyama, and Fumiko Itoh

CONTENTS

8.1 Introduction

Researchers in the humanities and social sciences, as shown in Part 3 of this volume, analyze many phenomena that are caused by, or related to, spatial factors. When the number of factors is small, spatial analysis with manual methods is tractable, but when the number is large, the analysis is laborious, and it often becomes intractable.

A few decades ago, researchers themselves used to develop computer programs to alleviate this task. However, the task required not only programming skills but also a lot of program-development time. As a result, the use of spatial analysis had very limited application to the humanities and social sciences. Nowadays, this difficulty has been overcome, largely by the introduction of Geographical Information Systems (GIS).

The ordinary GIS software provides many basic tools for spatial analysis (for example, "Spatial Analysts" in ArcGIS). However, when we wish to carry out advanced spatial analysis, the tools provided by ordinary GIS software are not always sufficient, and we have to find advanced ways. Fortunately, a considerable number of tools for advanced spatial analysis have been developed by the GIS community (Walker and Moor, 1988; Haslett et al., 1990; Openshaw et al., 1990; Openshaw et al., 1991; Okabe and Yoshikawa, 2003), and information about these tools is posted on the World Wide Web. Such information is, however, scattered over the Web, and it is difficult to find an appropriate tool for a specific spatial-analysis application. In fact, Google shows more than 3 million Web sites referring to "spatial analysis." The objective of this chapter is to introduce Web-based sites that are able to diminish this difficulty.

We first briefly introduce one of the most powerful search engines, served by the Center for Spatially Integrated Social Sciences (CSISS). Second, we show a Web-based system for finding free software packages for advanced spatial analysis, sited at the Center for Spatial Information Science (CSIS).

8.2   Search Engine at the CSISS Web Site

The CSISS Web site (www.csiss.org/search) provides five types of search engines:

1. Search for spatial resources.
2. Search the site.
3. Search social-science data archives.
4. Search for spatial tools.
5. Search of spatial-analysis literature in the social sciences.

All of these search engines are useful for studies in the humanities and social sciences, but the major concern of this chapter is with spatial tools, 4.

Clicking on 4 gives a dialog box, which asks us to enter a keyword for our specific spatial analysis; for example, "point pattern," in which case, 164 items will appear.

The information included in these items is classified into three types.

1. Description of methods for spatial analysis.
2. List of Web sites dealing with methods for spatial analysis.
3. Web sites providing software packages for spatial analysis.

The first type of information does not provide tools. The second type of information does not directly provide tools, but users may surf further to find a tool in the list. The last type of information does provide tools, but they may not be free. It is noted that users cannot specify the last type of information when they enter a keyword. Therefore, they have to examine 164 items to find an appropriate tool for their use. Professional spatial analysts can manage this task, but inexperienced or intermediate analysts may be overwhelmed by the huge amount of information. If they are particularly looking for free tools, much time is needed to find them. To overcome this difficulty, the Web system shown in the next section is developed.

8.3   FreeSAT: A Web System for Finding Free Spatial Analysis Tools

This section introduces *FreeSAT*, a system for finding Web sites that provide Free Spatial Analysis Tools, originally developed by Itoh and Okabe (2003). The address is ua.t.u-tokyo.ac.jp/okabelab/freesat/.

8.3.1   The home page of FreeSAT

The home page looks like this.

Welcome to **FreeSAT**: A Web system for finding Free Spatial Analysis Tools Version 2.0 developed by A. Masuyama, A. Okabe and F. Itoh

1. Spatial analysis for points
2. Spatial analysis for networks
3. Spatial analysis for attribute values of areas
4. Spatial analysis for continuous surfaces

FreeSAT classifies spatial analyses into four types: analysis for points, analysis for networks, analysis for attribute values of areas, and analysis for continuous surfaces. The first type of analysis deals with the distribution of point-like features, for example, the distribution of convenience stores in a region (Figure 8.1a). The second type of analysis deals with network-like features, for example, streets, railways, sewage, rivers, and so forth (Figure 8.1b). The third type of analysis deals with the attribute data of areas constituting a region; for example, population data by municipal districts (Figure 8.1c). The last type of analysis deals with an attribute value that is continuously distributed over a region, such as precipitation (Figure 8.1d). Users are required to choose the type of analysis suitable to their study.

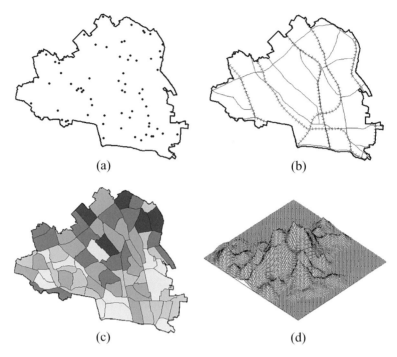

FIGURE 8.1
Examples of methods: (a) analysis for points, (b) analysis for networks, (c) analysis for attribute values of areas, and (d) analysis for continuous surfaces.

8.3.2 The "Spatial Analysis for Points" Page

Suppose that we want to analyze spatial patterns of point-like features (such as convenience stores in a city, as in Figure 8.1a). In this case, we click on "Spatial Analysis for Points" on the FreeSAT home page, and the following page appears.

1. SPATIAL ANALYSIS FOR POINTS

1.1 Point density estimation

1.2 Tests for clustered, random or dispersed

   1.2.1 Quadrat method

   1.2.2 Nearest neighbor distance method

   1.2.3 Ripley's K function and L-function

1.3 Detection of clusters

   1.3.1 Detection of spatial clusters

   1.3.2 Detection of spatio-temporal clusters

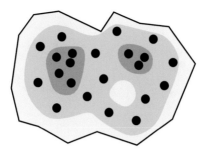

FIGURE 8.2
Point density estimation.

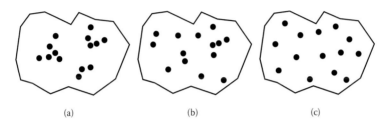

(a)   (b)   (c)

FIGURE 8.3
Point patterns: (a) clustered, (b) random, and (c) dispersed.

The methods are classified into three classes, namely "Point density estimation," "Test for clustered, random or dispersed," and "Detection of clusters." The first class (Section 1.1) deals with methods for estimating the density (indicated by the lightness of the gray color in Figure 8.2) from a given set of points (indicated by the points in Figure 8.2).

The second class of methods (Section 1.2) tests whether points are clustered (Figure 8.3a), random (Figure 8.3b), or dispersed (Figure 8.3c).

This test may be carried out using the "Quadrat," "Nearest neighbor distance," or "Ripley's K function and L function" method. The first method (Section 1.2.1) tests randomness in terms of the number of points in regularly shaped cells (e.g., squares) (Figure 8.4a). The second method (Section 1.2.2) tests randomness in terms of the distance from each point to its nearest point (Figure 8.4b). The third method (Section 1.2.3) tests randomness in terms of the cumulative number of points as a function of the distance from each point (Figure 8.4c).

The last class of methods (Section 1.3) detects clustered points in a plane (two-dimensional space) (Figure 8.5a) and in a spatio-temporal space (three-dimensional space) (Figure 8.5b).

8.3.3   The "Spatial Analysis for Networks" Page

Suppose that we next want to analyze network-like features, such as railways and roads, as in Figure 8.1b. In this case, we click on "Spatial

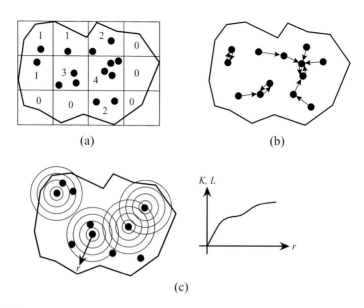

FIGURE 8.4
Tests for randomness: (a) the Quadrat method, (b) the nearest-neighbor distance method, and (c) the Ripley's *K*-function method.

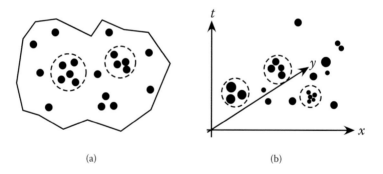

FIGURE 8.5
Detection of clusters in a plane (a) and in a spatio-temporal space (b).

Analysis for Networks" in the FreeSAT home page, and the following page appears.

2. SPATIAL ANALYSIS FOR NETWORKS

2.1 Topological analysis

2.1.1 Connectivity indices and accessibility indices

2.2 Network optimization

2.2.1 Shortest path problem

2.2.2 Maximum flow problem

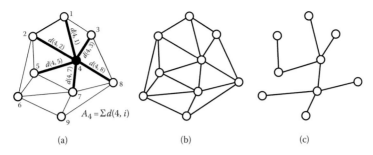

FIGURE 8.6
Accessibility index (a), and high connectivity (b) and low connectivity.

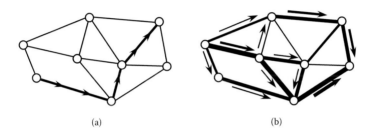

FIGURE 8.7
The shortest-path problem (a) and the maximum-flow problem (b).

"Topological analysis" (Section 2.1) deals with the topological nature of networks, such as accessibility indices (Figure 8.6a) and connectivity indices (Figure 8.6b, and 8.6c). "Network optimization" (Section 2.2) deals with two well-known problems, namely the shortest-path problem (Figure 7a) and the maximum-flow problem (Figure 7b).

8.3.4 The "Spatial Analysis for Attribute Values of Areas" Page

When attribute values (say, population) are given with respect to subregions (e.g., administrative districts) that constitute a whole study region (Figure 8.1c), and we want to analyze the distributional characteristics of these attribute values over that region, we click on "Spatial Analysis for Attribute Values of Areas" in the FreeSAT home page, and the following page appears.

3. SPATIAL ANALYSIS FOR ATTRIBUTE VALUES OF AREAS

3.1 Global spatial analysis

  3.1.1 Join-count statistics

  3.1.2 Spatial autocorrelation indices (Moran's I, Geary's C, Getis-Ord's G [d])

3.2 Local spatial analysis

  3.2.1 "Hot spots" detection

  3.2.2 Local spatial autocorrelation

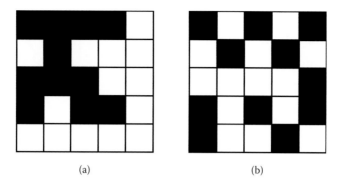

FIGURE 8.8
Join-count statistics, (a) associative and (b) dispersed.

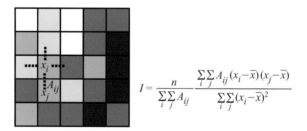

FIGURE 8.9
Spatial autocorrelation.

"Global spatial analysis" (Section 3.1) deals with the characteristics of the whole space, while "Local spatial analysis" (Section 3.2) deals with the characteristics of a local part of the whole space. The former analysis consists of two methods. The first method, i.e., the join-count statistics (Section 3.1.1), examines whether "black" cells tend to be spatially associative (Figure 8.8a) or dispersed (Figure 8.8b) in terms of the number of "B-B joins" and that of "B-W joins," where a "B-B join" means that two black cells are mutually adjacent.

The second method, i.e., spatial auto-correlation indices (Section 3.1.2), also examines whether or not similar values tend to be associative, but the values are continuous (gray color) in place of categorical values (black and white) (Figure 8.9).

"Local spatial analysis" (Section 3.2) is concerned with locally distinct places, often called "hot spots," in the whole space (Figure 8.10). Such places can be detected by the "hot spots" detection method (Section 3.2.1) or the local spatial-autocorrelation indices (Section 3.2.2).

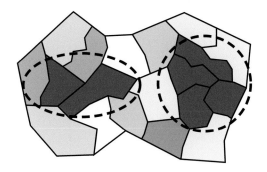

FIGURE 8.10
Detection of "hot spots."

8.3.5   The "Spatial Analysis for Continuous Surfaces Page"

This page looks like:

4. SPATIAL ANALYSIS FOR CONTINUOUS SURFACES

4.1 Estimation of a surface

    4.1.1 Spline interpolation

    4.1.2 Kriging method

    4.1.3 Trend surface analysis (polynomial fitting)

4.2 Topological surface network analysis

    4.2.1 Surface network analysis

    4.2.2 Contour tree analysis

This page deals with an attribute value continuously distributed over a region, which can be represented by a surface in three-dimensional space, such as precipitation over a region (Figures 8.1d and 8.11). In practice, the value is observable only at a finite number of points in the region (the points in Figure 8.11), and so we have to estimate the surface (the surface in Figure 8.11). In this case, we click on "Estimation of a surface" (Section 4.1), which includes the spline interpolation (Section 4.1.1), the kriging method (Section 4.1.2,) and the trend-surface analysis (Section 4.1.3).

Once a surface is estimated, we often want to analyze its qualitative (topological) characteristics. In this case, we click on "Topological surface network analysis" (Section 4.2), which includes two methods. Both surface-network analysis (Section 4.2.1) and contour tree analysis describe the topological characteristics of a surface in terms of the configuration of "peaks," "col," and "bottoms," (Figure 8.12). They vary, in that the rules for joining these critical points (the continuous lines in Figure 8.12) are different.

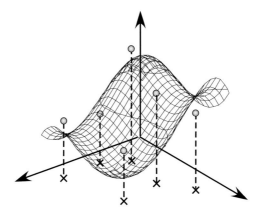

FIGURE 8.11
Estimation of a surface.

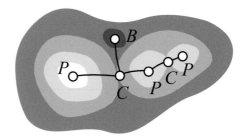

FIGURE 8.12
Topological surface-network analysis.

8.3.6   Tables of Software Names

When we find an appropriate method, for example, "point density estimation," we click on that method, and a table, such as Table 8.1, appears. This shows the names of free software packages that include "point density estimation." We notice from this table that ANTELOPE, CrimStat, Field, GRASS and HOTSPOT provide free software packages for point-density estimation. If we click on one of the names, then we jump to the Web site providing this software package. Following the instruction given there, we can obtain a free software package. Similar tables are also given with respect to "Spatial Analysis for Networks," "Spatial Analysis for Attribute Values of Areas," and "Spatial Analysis for Continuous Surfaces."

TABLE 8.1

The Names of Free Software Packages with Respect to the Methods of Spatial-Point Analysis

Software	1. Analysis for Points					
		1.2 Tests for Clustered/Random/Dispersed			1.3 Detection of Clusters	
	1.1 Point-Density Estimation	1.2.1 Quadrat Method	1.2.2 Nearest Neighbor Distance Method	1.2.3 Ripley's *L* Function *K* Function	1.3.1 Detection of Spatial Clusters	1.3.2 Detection of Spatio-Temporal Clusters
ANTELOPE	@					
Cluster					@	@
Clustering Calculator					@	
CrimeStat	@		@	@	@	
Field	@					
FRAGSTATS			@			
GAM, GCEM, GEM					@	
GMT			@			
GRASS	@					
HOTSPOT	@					
IDRISI		@				
MOVEMENT			@			
NEM					@	
Pointstat			@			
Potemkin				@		
PPA			@	@		@
R Package					@	
S + Modern Applied Statistics				@		
SADA			@			
Spatial Statistics Toolbox			@			
SPATSTAT			@	@		
Spheri Stat	@					
Splancs	@		@	@		
SPPA				@		

8.4   Conclusion

As shown in the preceding sections, FreeSAT is a Web system for searching for free tools for spatial analysis on the Web space. The search may be initiated by answering the following questions.

"What types of features does your study deal with: points, networks, areas, or surfaces?"

In the case of points:

"Do you want to estimate the density of points?" (Yes, then visit the page of Section 1.1).

"Do you want to test whether points are clustered, random, or dispersed?" (Yes, then visit the page of Section 1.2).

"Do you want to detect clustered points?" (Yes, then visit the page of Section 1.3).

In the case of networks:

"Do you want to measure connectivity or accessibility?" (Yes, then visit the page of Section 2.1).

"Do you want to find the shortest path or the maximum flow?" (Yes, then visit the page of Section 2.2).

In the case of areas:

"Do you want to analyze the global characteristics of attribute values over the areas?" (Yes, then visit the page of Section 3.1).

"Do you want to analyze the local characteristics of attribute values over the areas?" (Yes, then visit the page of Section 3.2).

In the case of surfaces:

"Do you want to estimate a surface from the values at points?" (Yes, then visit the page of Section 4.1).

"Do you want to analyze qualitative characteristics of a surface?" (Yes, then visit the page of Section 4.2).

We hope that FreeSAT helps you find the appropriate free tool that you are looking for.

References

Getis, A., Spatial analysis and GIS: an introduction, *J. Geogr. Syst.*, 2, 1–3, 2000.

Itoh, F. and Okabe, A., A Web System for Finding Free Software of Spatial Analysis (Abstract), the annual meeting of the Association of American Geographers, New Orleans, 2003.

Okabe, A. and Yoshikawa, T., SAINF: A toolbox for analyzing the effect of point-like, line-like and polygon-like infrastructural features on the distribution of point-like non-infrastructural features, *J. Geogr. Syst.*, 5, 407–413, 2003.

Openshaw, S., Cross, A., and Charlton, M., Building a prototype geographical correlates exploration machine, *Int. J. Geogr. Info. Sys.*, 4, 297–311, 1990.

Openshaw, S., Brunsdon, C., and Charlton, M., A spatial analysis toolkit for GIS, European Conference on Geographical Information Systems, 788–796, 1991.

Haslett, J., Wills, G., and Unwin, A., SPIDER-an interactive statistical tool for the analysis of spatially distributed data, *Int. J. Geogr. Info. Sys.*, 4, 285–296, 1990.

Walker, P.A. and Moore, D.M., SIMPLE: an inductive modeling and mapping tool for spatially-oriented data, *Int. J. Geogr. Info. Sys.*, 2, 347–363, 1988.

9

A Toolbox for Examining the Effect of Infrastructural Features on the Distribution of Spatial Events

Atsuyuki Okabe and Tohru Yoshikawa

CONTENTS

9.1 Introduction

In the real world, there are many events that occur at specific locations. These are called *spatial events,* and they include the location of facilities in particular places. Spatial events are in part affected by their constraining geography, in particular by influencing elements that persist over a long time period. These durable controls are called *infrastructural features.* Examples of these that have attracted research in the humanities and social sciences are as follows:

- Transport stations attract crime in Los Angeles (Loukaitou-Sideris et al., 2002).

- Mosques are usually located on hilltops in Istanbul (Kitagawa et al., 2004).

- Steel mills are distant from their supportive mines when considering the period from 1974 to 1991 in the United States (Beeson and Giarratani, 1998).

- Asthma sufferers reside 200–500 meters from major highways in Erie County, New York (Lin et al., 2002).

- Serial thieves in Baltimore have a tendency to migrate south along the major roads (Harries, 1999).

- Early ceramic sites, especially those yielding fiber-temper pottery, had been found along the coast or close to mangrove stands in Ecuador (Marcos, 2003).

- Luxury apartment buildings are preferentially located around big parks in Setagaya, Tokyo (Okabe et al., 1988).

This chapter introduces a user-friendly toolbox, called SAINF (Okabe and Yoshikawa, 2003), which may be used in the statistical analysis of these spatial relationships. SAINF is the abbreviated name for Spatial Analysis of the Effect of Infrastructural Features.

9.2   General Setting

We consider a region where spatial events occur, and within which infrastructural features are placed. Such infrastructural features have various geometrical forms that can be classified into three types: point-like, such as railway stations; line-like, seen as roads; and polygon-like, exemplified by city parks. It should be noted that this classification is relative, in the sense that a station may be a polygon on a large-scale map but a point at small scale.

In the Geographical Information Systems (GIS) environment to which SAINF is applied, geographical features and spatial events are represented by geometrical objects that are points, line segments, and polygons. Spatial events are points on a plane. The number m, of infrastructural features, for example, railway stations, is denoted by $o_1, \ldots, o_m$ and the number of spatial events n, such as crime locations is given by $p_1, \ldots, p_n$. We assume that the spatial events do not occur on the infrastructural sites; that is, points $p_1, \ldots, p_n$ are placed on the complement of the area O, occupied by $o_1, \ldots, o_m$ with respect to a study region S, that is, $S \setminus O$.

SAINF statistically tests the following hypothesis, H_o, to examine the effect of the configuration of infrastructural features $o_1, \ldots, o_m$ on the distribution of spatial events $p_1, \ldots, p_n$,

Ho: Spatial events $p_1, \ldots, p_n$ occur uniformly and randomly over the region.

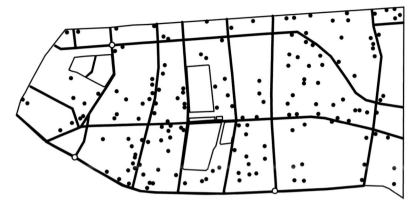

FIGURE 9.1
Railway stations (white circles), streets, big parks, and luxury apartment buildings (black circles) in Kohtoh, Tokyo.

In geometrical terms, points p_1, , p_n are uniformly and randomly distributed over the region $S \setminus C$.

When related to H_o, "uniformly and randomly" implies that spatial events are distributed independently of the configuration of infrastructural features. If this hypothesis is rejected, the infrastructure may have an effect.

9.3 Procedure for Examining the Effect

We consider an example of how to use SAINF by examining the influence of three infrastructural elements on the location of luxury apartment buildings in Kohtoh, Tokyo. These are seen as black circles in Figure 9.1. The three factors are railway stations, arterial streets, and big parks, given by white circles, line segments, and polygons, respectively, in Figure 9.1.

Data concerning stations, streets, parks, and apartment buildings may be available in the form of digital or paper maps. If the latter, we digitize the geographical features.

GIS software varies, but SAINF adopts ArcView as one of the most popular GIS viewers. This system employs the "shapefile" format specific to ArcView.

To use SAINF, we install the software package ArcView together with that of SAINF. The latter may be downloaded without charge for nonprofit-making uses from the Web site: ua.t.u-tokyo.ac.jp/okabelab/atsu/sainf/.

ArcView software is available at cost from Environmental Systems Research Institute, Inc. (ESRI).

Once SAINF, ArcView and the datasets are installed, we are ready to start the analysis. Clicking on "SAINF-Tools" on the ArcView menu bar reveals a menu showing the available tools. SAINF provides three tools, which are the goodness-of-fit test, conditional nearest-neighbor distance, and the cross K function methods. The goodness-of-fit test method is first considered.

9.3.1   The Procedure for Using the Goodness-of-Fit Test Method

The goodness-of-fit test method of SAINF generally tests the hypothesis H_o by comparing the observed number of point spatial events for each subregion with the expected point numbers that would be realized under a condition in which spatial events are uniformly and randomly distributed, as envisaged by hypothesis H_o.

Subregions are considered to be "buffer rings" for the infrastructural features. A *buffer ring*, $R(d_i, d_{i+1})$ is the region in which the distance to its nearest infrastructural feature is between d_i and d_{i+1} ($d_i < d_{i+1}$). The boundaries of buffer rings are equidistant contour lines around infrastructure elements, examples of which are shown in Figures 9.2a, 9.2b, and 9.2c.

We use one of the functions of ArcView to generate buffer rings. In the dialog box of "Buffer Wizard," a set of infrastructural features is chosen in the pull-down menu "The features of a layer," for example, railway stations, and the number k of buffer rings and the width of a ring $d_{i+1} - d_i$ are entered. We also enter the name of an output file for the result. After a few seconds of computation, the buffer-contour rings appear, as shown in Figure 9.3.

The buffer rings cover the study region, which is the polygon in Figure 9.3. To trim them outside of the study region, we use the "Geoprocessing Wizard." Appropriate names or items are chosen in "Clip one layer based on another," "Select the input layer to clip," "Specify a polygon clip layer," and "Specify the output shapefile or feature class." Trimmed buffer rings are achieved, as seen in Figure 9.2a.

Since the goodness-of-fit test method tests the hypothesis H_o, and, while recalling that spatial events are uniformly and randomly distributed over a region, we notice that the number of events occurring in a subregion is proportional to the area of the subregion. The hypothesis H_o can therefore be restated as follows:

H'_o: The number of spatial events occurring in a buffer ring $R(d_i, d_{i+1})$ is proportional to the area of R (d_i, d_{i+1}). In geometrical terms, the number of points that are placed in $R(d_i, d_{i+1})$ is proportional to the area of $R(d_i, d_{i+1})$.

To test this hypothesis, we have to measure the area of $R(d_i, d_{i+1})$. Clicking on "Areas of buffer rings" in the "SAINF-Tools" menu, the dialog box appears, where we enter the names of the buffer layer, an input file showing ring intervals, and an output file for the results. After a few seconds of computation, SAINF produces a display, such as the one shown in Table 9.1. From this result, we obtain the ratio P_i for the area of each buffer ring $R(d_i, d_{i+1})$ to the total area. Since the total number of points is n, the *expected* number of points that would be placed in a buffer ring $R(d_i, d_{i+1})$ under the null hypothesis H_o is $n\, P_i$.

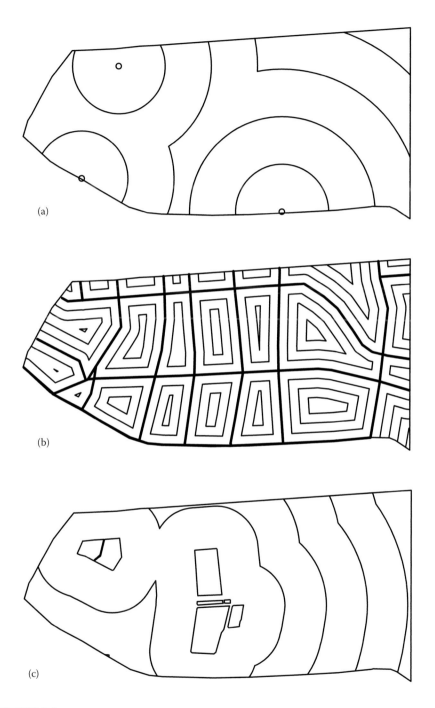

FIGURE 9.2
The buffer rings of (a) railway stations, (b) arterial streets, and (c) big parks.

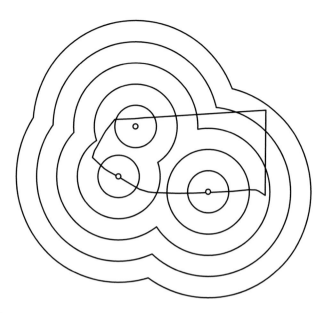

FIGURE 9.3
The untrimmed buffer rings of railway stations.

TABLE 9.1

The Area of Each Buffer Ring

i	d_{i-1}	d_i	$R(d_{i-1}, d_i)$
1	0	450	1,120,512
2	450	900	2,155,189
3	900	1350	1,500,716
4	1350	1800	676,378
5	1800	2250	124,281

Our next task is to count the *observed* number of points placed in $R(d_i, d_{i+1})$. We click on "Number of non-infra features" in the "SAINF-Tools" menu, and a dialog box appears, showing the layer of noninfrastructural features. "Luxury apartment buildings" is selected. A second dialog box appears, showing the layer of study regions. "Kohtoh" is entered. A third dialog box appears, asking the name of an input file showing ring intervals and of an output file for the result. We put in the names of these file, and in a few seconds, a display, such as the one in Table 9.2, is shown.

As stated in the hypothesis H'_o, if the hypothesis H_o was valid, the observed number $\bar{N}_i$ would be proportional to the area of $R(d_i, d_{i+1})$. In other words $\bar{N}_i$ would be close to the expected number nP_i for $i = 1, \quad ,k$ where k is the number of rings (i.e., $\bar{N}_i - nP_i \approx 0$). Therefore, the value of χ^2 as defined by:

TABLE 9.2

The Number of Luxury Apartment
Buildings in Each Buffer Ring

i	d_{i-1}	d_i	N_i
1	0	450	24
2	450	900	68
3	900	1350	45
4	1350	1800	19
5	1800	2250	6

$$\chi^2 = \sum_{i=1}^{k} \frac{\left(\bar{N}_i - nP_i\right)^2}{nP_i} \tag{9.1}$$

would be "significantly" small if the null hypothesis H_o held. Conversely, if the hypothesis H_o does not stand, this value would be "significantly" large. The statistical theory of the goodness-of-fit test (Peason, 1900) gives critical values for "significantly" large, and these are tabulated in a chi-square distribution table. We can test the validity of hypothesis H_o by consulting this table.

In our example, the observed numbers $\bar{N}_i$ and the expected numbers $n\,P_i$ with respect to buffer rings $i = 1, \quad ,5$ are shown in the third and fourth columns of Table 9.2. The "chi-square probability" in this case is 0.360, implying that the probability of the observed numbers being realized under the hypothesis H_o is 0.360. This is larger than the 0.05 significance level. We can, therefore, conclude that the configuration of railway stations has no effect on the distribution of luxury apartment buildings in Kohtoh.

9.3.2 The Procedure for Using the Conditional Nearest-Neighbor Distance Method

A second tool in SAINF is the conditional nearest-neighbor distance method proposed by Okabe and Miki (1984) and Okabe et al. (1988) and Okable an Yoshikawe (1989), which is a modification of the *nearest neighbor distance*, or *NN distance* method (Dacey, 1968). This procedure is based on the theory that the observed average distance $\bar{d}$ from each point p_i to its nearest infrastructure $(i = 1,\ldots,n)$ would be "significantly" shorter or "significantly" longer than the expected average NN distance, μ that would be obtained under the null hypothesis H_o.

If the hypothesis H_o is false, the absolute value of z defined by

$$z = \frac{\bar{d} - \mu}{\sigma / \sqrt{n}} \tag{9.2}$$

is "significantly" large or "significantly" small, where σ is the standard deviation of the NN distance under the hypothesis H_o.

The central-limit theorem (Gnedenko and Kolomogorov, 1954) shows that the value z follows the standard normal distribution when n is a large number, and the critical values for the "significantly" large or small values are obtained from a statistical table for standard normal distribution. We can test the hypothesis by consulting this table.

To perform the method above using SAINF, the menu "SAINF-Tools" is opened, and "NN distance" is clicked on. A first dialog box, "Infra feature layer" appears, showing the types of infrastructural features. We click on "arterial streets." A second dialog box, "Non-infra feature layer," appears giving the names of noninfrastructural features. "Apartment buildings" is selected. A third box, "Study region layer" appears, and "Kohtoh" is chosen. A fourth dialog box is then revealed, "Do you want to visualize the lines from non-infra features to their nearest infra features?" If we click on "yes," Figure 9.4 is shown. Each line segment traces the distance from each luxury apartment building to its nearest point on the street network.

SAINF measures the distance from each point to its nearest approach to the arterial streets. To obtain the observed average NN distance, we click on "Average NN distance" in the "SAINF-Tools" menu. In the dialog box, "Non-infra structural layer," "apartment buildings" is selected. After a few seconds, SAINF shows the observed average NN distance $\bar{d}$ and the number n of apartment buildings. These are: $\bar{d} = 91.36$ m and $n = 162$.

To obtain the expected NN distance that would be realized under the hypothesis H_o, we click on "Mean & standard deviation" on the "SAINF-Tools" menu. After a brief delay, SAINF gives $\mu = 94.973$ and $\sigma = 70.85$. Insertion of these values into Equation (9.2) obtains $z = -0.649$. The standard, normal distribution table shows that the critical values are less than -1.96

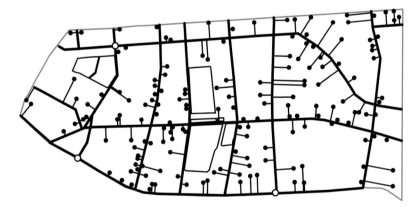

FIGURE 9.4
Line segments indicating the distance from each luxury apartment building to its nearest point on the arterial streets in Kohtoh, Tokyo.

or greater than 1.96, which is smaller than the z value. We may therefore conclude that the configuration of arterial streets has no effect on the distribution of luxury apartment buildings in Kohtoh.

9.3.3   The Cross K Function Method

A third tool within SAINF is the cross K function method (Ripley, 1981). This is similar to the conditional NN distance method where infrastructural features are considered as point-like features. A difference exists in that the latter considers the distance from each event to its nearest infrastructural feature, but the former considers the distance from each infrastructural feature to all spatial events, thereby considering the more global aspects.

To state the cross K function precisely, we consider a point-like infrastructural feature o_i and the number $K_i(t)$ of spatial event points that are located within distance t from the infrastructure o_i, where t is a variable. An illustration of $K_i(t)$, $i = 1$ is shown in Figure 9.5.

We notice from panel (a) that the number of points in the circle centered at p_1 with radius $t = 1$ is two, and so $K_1(1) = 2$. The number of points in the circle with radius $t = 2$ is five, and so $K_1(2) = 5$, and so forth. As a result, the function $K_1(t)$ as in panel (b) is obtained. Similarly, we derive functions $K_i(t)$ for p_i, $i = 2, ..., m$. Averaging the resulting functions, $K(t)$ emerges as:

$$K(t) = \frac{1}{m} \sum_{i=1}^{m} K_i(t). \tag{9.3}$$

The function $K(t)$ is called the *cross K function*. We consider two cross K functions. The first is an *observed* cross K function, which is obtained for the actual distribution of spatial-event points, such as luxury apartment buildings, shown in Figure 9.1. The second is an *expected* cross K function, which is obtained considering the null hypothesis H_o, which assumes that spatial

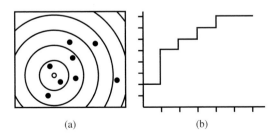

(a)   (b)

FIGURE 9.5
Function $K_1(t)$.

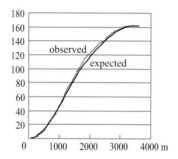

FIGURE 9.6

The observed cross K function and the expected cross K function for railway stations and luxury apartment buildings in Kohtoh, Tokyo.

events are uniformly and randomly distributed over a region. If the observed cross K function is similar to the expected cross K function over possible values of t, we conclude that spatial events tend to be independent of the configuration of the point-like infrastructural features.

We can achieve the cross K function method by selecting "Cross K function" and "Expected Cross K function" from the "SAINF-Tools" menu. Since the procedure is similar to the conditional NN distance method seen in Section 9.3.1, the procedure is not shown here, but the result may be seen in Figure 9.6. This figure shows that the observed cross K -function is almost the same as the expected cross K function, although the former is slightly greater than the latter, around 1800–2500 meters. We may conclude from this that railway stations do not have a major influence on the distribution of luxury-class apartment buildings.

9.4 Conclusion

This chapter introduced SAINF, a GIS-based toolbox that is designed to examine the effect of the configuration of infrastructural features on the distribution of spatial events. A distinctive characteristic of SAINF is that it can be applied to point-like infrastructural elements, as well as line-like and polygon-like features. This chapter outlines the procedure for operating SAINF. If the reader wishes to know more details he/she should consult with the SAINF manual, which can be downloaded from the Web site noted in Section 9.1. We anticipate that SAINF will help those scholars undertaking research in the humanities and social sciences who wish to understand the underlying factors of spatial phenomena.

Acknowledgments

We express our thanks to Exceed Co. Ltd. for helping us program SAINF, and to Miki Arimoto for deriving the data and figures of parks and luxury apartment buildings in Kohtoh. This development was partly supported by Grant-in-aid for Scientific Research No. 10202201 and No. 14350327 of the Ministry of Education, Culture, Sports, Science and Technology, Japan.

References

Beeson P. and Giarratani, F., Spatial aspects of capacity change by U.S. integrated steel producers, *J. Reg. Sci.*, 38, 425–444, 1998.

Dacey, M.F., Two-Dimensional Random Point Patterns, A Review and an Interpretation, *Papers of the Regional Science Association*, Vol.13, 1968, 41–55.

Gnedenko, B.V. and Kolomogorov, A.N., *Limit Distribution for Sum of Independent Random Variables* (translated from the 1949 Russian ed., Chung, K.L., Ed.), Addison-Wesley, Reading, 1954.

Harries, K., Mapping Crime: Principle and Practice, Research Report, National Institute of Justice, 1999.

Kitagawa, K., Asami, Y. and Neslihan, D. Three dimenstional view analysis using GIS: the locational tendency of mosques in Bursa, Turky in *Islamic Area Studies with Geographical Information Systems*, Okabe, A., Eds., Routledge Curson, London, 243–252, 2004.

Lin, S., Munsie, P.J., Hwang, S-A., Fitzgerald, E., and Cayo, M.R., Childhood asthma hospitalization and residential exposure to state route traffic, *Environ. Res.*, 88, 73–81, 2002.

Loukaitou-Sideris, A., Liggett, R., and Iseki, H., The geography of transit crime: documentation and evaluation of crime incidence on and around the Green Line Stations in Los Angeles, *J. Plann. Educ. Res.*, 22, 135–151, 2002.

Marcos, J.M., A reassessment of the Ecuadorian formative, in *Archaeology of Formative Ecuador*, Raymond, J.S. and Burger, R.L., Eds., Dumbarton Oaks Research Library and Collection, Washington, D.C., 7–32, 2003.

Okabe, A. and Miki, F., A conditional nearest-neighbor spatial association measure for the analysis of conditional locational interdependence, *Environ. Plann. A*, 16, 163–171, 1984.

Okabe, A., Fujii, A., Oikawa, K., and Yoshikawa, T., The statistical analysis of a distribution of activity points in relation to surface-like elements, *Environ. Plann. A*, 20, 609–620, 1988.

Okabe, A. and Yoshikawa, T., Multi nearest distance method for analyzing the compound effect of infrastructural elements on the distribution of activity points, *Geogr. Anal.*, 21, 216–235, 1989.

Okabe, A. and Yoshikawa, T., SAINF: a toolbox for analyzing the effect of point-like, line-like and polygon-like infrastructural features on the distribution of point-like non-infrastructural features, *J. Geogr. Sys.*, (5), 407–413, 2003.

Peason, K., On a criterion that a system of deviations from the probable in the case of a correlated system of variables is such that it can be reasonably supposed to have arisen in random sampling, *Philosophical Magazine*, 50, 157–175, 1900.
Ripley, B.D., *Spatial Statistics*, John Wiley, Chichester, 1981.

10

A Toolbox for Spatial Analysis on a Network

Atsuyuki Okabe, Kei-ichi Okunuki, and Shino Shiode

CONTENTS

10.1 Introduction

In the real world, we notice many events and situations that locate at specific points on a network. These are referred to as *network spatial events*. Some typical examples relevant to studies in the humanities and social sciences are as follows:

Homeless people living on the streets (Arapoglou, 2004).

Street crime (Harries, 1999; Painter, 1994; Ratcliffe, 2002).

Graffiti sites along streets (Bandaranaike, 2003).

Urban cholera transmission (Snow, 1855).

Traffic accidents (Yamada and Thill, 2004).

Illegal parking (Cope, 1990).

Street food stalls (Stavric, 1995; Tinker, 1997).

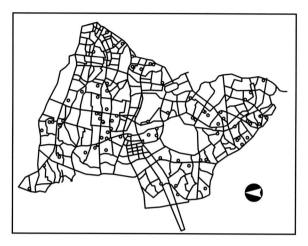

FIGURE 10.1
Churches alongside the streets in Shibuya-Shinjuku, Tokyo.

In addition to the types of events listed above, there is another large class also representing network spatial events, but these occur alongside a network. A typical example is shown in Figure 10.1, where the circles indicate the locations of churches in Shibuya-Shinjuku, Tokyo. It can be seen that these are not freely situated over the region, since their positions are strongly constrained by their location along the streets.

Not only churches, but also almost all facilities in an urbanized area, are located at the side of streets, and it is actually the gates or entrances of these facilities that lie adjacent to the thoroughfare.

This chapter focuses on the analysis of events and facilities that are placed at specific locations on and alongside a network, and are called *network spatial events*.

A decade ago, analysis of network spatial events was very difficult, because network data were poor and there were few tools for their analysis, such that researchers had to assemble data and develop methods themselves. This task demanded much time and effort. The modern advent of geographical information systems (GIS) and the abundance of network data that are accessible today have, fortunately, made matters easier, and many GIS-based tools are available. In this chapter, we introduce a user-friendly toolbox, called SANET, which is the abbreviated name for Spatial Analysis on a Network. This tool is useful for answering, for instance, the following questions:

Does illegal parking tend to occur uniformly in no-parking streets?

Are street crime locations clustered in "hot spots"?

Do fast-food shops tend to contend with each other?

How extensive is the service area of a post office?

What is the probability of consumers choosing a particular downtown store?

In the subsequent sections, we show how to answer these questions using SANET.

10.2 Tools in SANET

SANET was released in November 2001, and it has been evolving ever since (Okabe, Okunuki, and Shiode, 2004). The current 2005 edition of SANET is the third version, and it provides the following 15 tools:

1. Construction of a node-adjacency data set.
2. Assignment of a data point to the nearest point on a network.
3. Aggregation of attribute values belonging to the same item.
4. Generation of a network Voronoi diagram.
5. Generation of random points on a network.
6. Enactment of the network cross K function method.
7. Enactment of the network K function method.
8. Partition of a polyline into constituent line segments.
9. Assignment of polygon attributes to the nearest line segment.
10. Enactment of the nearest-neighbor distance method.
11. Enactment of the conditional nearest-neighbor distance method.
12. Calculation of polygon centroids.
13. Enactment of the network Huff model.
14. Enactment of the variable clumping method.
15. Comparison of two networks.

In the subsequent sections, the procedure for spatial analysis on a network using these tools is outlined. First, in Section 10.3, SANET and datasets set up on the computer are described. Second, in Sections 10.4–10.8, we show how to achieve spatial analysis with the network K function method (Tool 7) using an illustrative example in Figure 10.1; also shown are the network variable-clumping method (Tool 14), the network cross K function method (Tool 6), the network Voronoi diagram (Tool 4), and the network Huff model (Tool 13).

10.3 Software and Data Setting

The software SANET consists of two components: the main program, and the interface between this and a GIS viewer.

The main program performs the geometric and algebraic computation needed for running the tools mentioned in Section 2. This program works independently, and can, in theory, be interfaced with any GIS viewer. The interface between the main program and a viewer will clearly depend on the choice made from the many viewers available. SANET currently adopts ArcView, which is one of the most popular GIS viewers. The main program and the interface can be downloaded from the SANET Web site: http://okabe.t.u-tokyo.ac.jp/okabelab/atsu/sanet/sanet-index.html.

This download can be made without charge for nonprofit-making uses. Also posted on this Web site is the detailed manual of SANET and information about the most recent version. The GIS viewer ArcView is obtainable at a reasonable price from Environmental Systems Research Institute, Inc. (ESRI).

After installing both SANET and ArcView on a personal computer, the computer-readable digital data of a street network and churches has to be obtained. There are many ways of recording and managing the digital data of a street network. The main program of SANET employs *adjacent-node tables* that are commonly used in computational geometry. The adjacent-node tables for the street network of Figure 10.2 are shown in Table 10.1. This illustration consists of straight-line segments whose end points (called *nodes*) are labeled by numbers. Table 10.1(a), called a *header table*, shows that node *i*, say node 0, is headed to the ID = 0 in Table 10.1(b). Table 10.1(b) shows that the nodes adjacent to the node corresponding to ID = 0 (i.e., node 0) are nodes 1 and 5 (reading downwards).

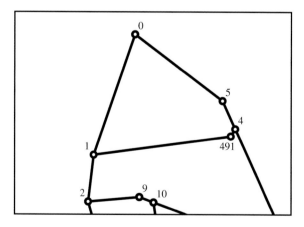

FIGURE 10.2
Nodes of a street network.

TABLE 10.1

Adjacent Node Tables

(a) Header table		(b) Adjacent node table	
Node ID	Head	ID	Adjacent Node
0	0	0	1
1	2	1	5
2	5	2	0
3	8	3	2
4	11	4	491

The structure of street data varies in differing GIS software packages. ArcView uses Polyline, which is not compatible with the adjacent-node tables. Therefore, when SANET is used, we have to transform Polyline to enable it to function. This transformation is made by using Tool 1.

The digital data for churches may be given either as the coordinates of their representative centroid points or as polygons representing the areas occupied by the buildings. SANET assumes that features are represented by points. With data given in the latter form, the centroids of the polygons are easily located by using Tool 12.

For SANET, all network spatial events are precisely on a network. As is seen in Figure 10.1, churches are not exactly located on streets, because a point does not indicate the gate of a church but the centroid of its buildings. In practice, these entrance data are difficult to obtain, and, hence, we have to estimate them from the centroids. SANET assumes that the nearest point on a street from the centroid of a facility is its gate. The location of these *access points* is derived by using Tool 2. An example is given in Figure 10.3, which shows the access points of the churches plotted in Figure 10.1.

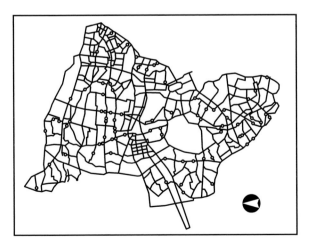

FIGURE 10.3
The access points of churches in Shibuya-Shinjuku, Tokyo, obtained using Tool 2.

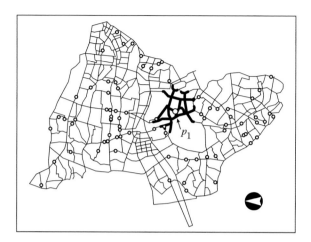

FIGURE 10.4

The sub-network in which the distance from p_1 is less than or equal to 1000 meters.

We are ready to analyze network spatial events, now that SANET and the data are set up.

10.4 Network *K* Function Method

When observing the distribution of churches in Shibuya-Shinjuku, seen in Figure 10.3, we wonder whether they are clustered, random, or dispersed. There are many methods available for this analysis, and SANET provides two tools to enable the determination to be performed. These methods are the *K*-function (Tool 7), and the nearest-neighbor distance (Tool 10). The first of these approaches is used below.

The *K*-function method was originally formulated on a plane by Ripley (1981), and this was extended by Okabe and Yamada (2001) to apply to a network. The *K*-function is formulated in terms of the function $K_i(t)$ defined as the cumulative number of points representing events within the shortest-path distance t, from a point, p_i, $i = 1, \ldots n$, where n is the number of points. For example, the bold lines in Figure 10.4 indicate the sub-network in which the distance from p_1 is less than or equal to 1000 meters. Since two churches, represented by the two circles on the bold lines, are located on this sub-network, the value of $K_1(t)$ for $t = 1000$ meters is two, i.e., $K_1(1000) = 2$. By extending t from 0 to 7000, we obtain the function $K_1(t)$ as in Figure 10.5. In terms of $K_i(t)$, the *K*-function, $K(t)$, is written as:

$$K(t) = \frac{1}{n} \sum_{i=1}^{n} K_i(t) \qquad (10.1)$$

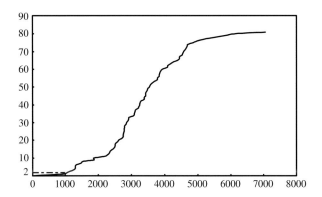

FIGURE 10.5
$K_1(t)$ function for the church at p_1 in Figure 10.4.

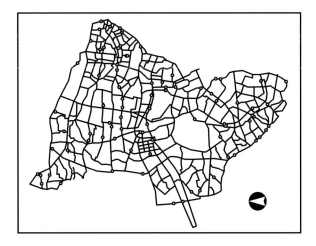

FIGURE 10.6
Randomly and uniformly generated points on the streets in Shibuya-Shinjuku, Tokyo, using Tool 5 (the number of points is the same as that in Figure 10.1).

This implies that the K function is the average of the functions $K_i(t)$ across $i = 1... n$.

To examine whether the churches tend to be clustered or dispersed, the *observed* K-function, which is obtained from given data, is compared with the *expected* K-function obtained when spatial-event points are uniformly and randomly distributed over the network. Figure 10.6 shows such a real-ized set of points for the streets in Shibuya-Shinjuku using Tool 5 (the number of points is the same as that in Figure 10.3). To obtain the expected K function, as many as 1000 sets of points are generated, and the resulting K functions are averaged to give the approximate expected result.

Figure 10.7 shows the observed K function and the expected K function for the churches in Shibuya-Shinjuku, obtained by using Tool 7. The observed K function (the black curve) is always above the expected K function (the

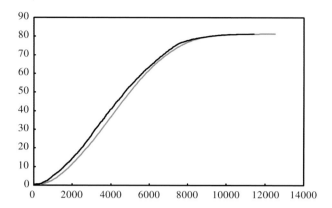

FIGURE 10.7
The observed (the black curve) and expected (the gray curve) *K* functions for churches in Shibuya-Shinjuku, Tokyo.

gray curve). This implies that the churches in Shibuya-Shinjuku tend to be clustered rather than randomly distributed.

10.5 Network Variable-Clumping Method

The finding in Section 10.4 suggests that there may be a distinct pattern of "clumps" in the distribution of churches in Shibuya-Shinjuku. To examine whether or not such a pattern exists, the variable-clumping method (Tool 14) is employed. The clumping method on a plane, originally devised by Roach (1968), was developed into the variable-clumping method on a plane by Okabe and Funamoto (2000), and extended to a network by Shiode and Okabe (2004).

To explain the meaning of a "clump," we define the *r-neighborhood* of a point, p_i , as the sub-network of a network in which the shortest-path distance from p_i to any point in the *r*-neighborhood is less than or equal to *r*, which is called the *clump radius*. The *r*-neighborhood of p_1 is indicated by the bold, gray line in Figure 10.8. A *clump* is a set of points whose *r*-neighborhoods form one connected sub-network of a network (Figure 10.8). The number of points forming a clump are referred to as the *clump size*. This varies from 1, (one point forms one "clump"); to *n* (all the points form one clump). The state of clumping, called the *clump state* and denoted by $C(r)$, is described in terms of the number, $N(k \mid r)$, of clumps with respect to clump size, *k*, that is:

$$C(r) = (N(1 \mid r), N(2 \mid r), \quad , N(n \mid r)) \qquad (10.2)$$

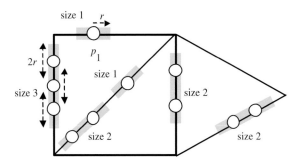

FIGURE 10.8
Clumps with radius *r*.

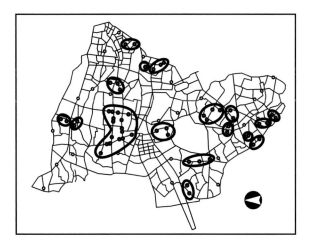

FIGURE 10.9
A significant clump state observed in the distribution of churches in Shibuya-Shinjuku, Tokyo.

Since $N(1\,|\,r) = 2, N(2\,|\,r) = 3, N(3\,|\,r) = 1, N(i\,|\,r) = 0, i = 4,\ldots,11$ in Figure 10.8, the clump state of the points in Figure 10.8 is $C(r) = (2, 3, 1, 0, 0, 0, 0, 0, 0, 0, 0)$.

In the above, the clump radius *r* is constant, but it can be variable. The method of describing the clump state from the smallest to the largest value of *r* is called the *variable-clumping method*. In this way, local clumps, and also global clumps, can be shown.

Among many possible clump states, there is a need to detect "significant" ones. A *significant clump state* is defined as that one which rarely occurs (i.e., occurs with small probability) in the context of points being uniformly and randomly distributed over a network. The significant clump states can be discerned by generating random points many times using Tool 5. Figure 10.9 shows one of the significant clump states observed in the distribution of churches in Shibuya-Shinjuku, which was detected by using Tool 14. The clump radius is 500 meters, and the probability of realizing a pattern like

Figure 10.9 is less than 0.05. This significant clump state is characterized by one clump of size 18, one clump of size 5, five clumps of size 3, 10 clumps of size 2, and 24 clumps of size 1; in particular, the clump of size 18 is distinctive.

10.6 Network Cross K Function Method

The observation in Section 10.4 may suggest that the churches tend to be located around transport stations. This trend can be examined by the network cross K function method (Tool 6) formulated by Okabe and Yamada (2002), which is an extension of the cross K function method defined on a plane.

The cross K function method is similar to the K function method mentioned in Section 10.3.2, but the root points are different. The sets of points considered are those of churches, $p_1, \ldots, p_n$, and of stations, $q_1, \ldots, q_m$. A function, $K_i^C(t)$, is defined as the cumulative number of churches within the shortest-path distance, t, from a station, q_i, $i = 1 \ldots m$, where m is the number of stations. The cross K function is defined by:

$$K^C(t) = \frac{1}{m} \sum_{i=1}^{m} K_i^C(t) \tag{10.3}$$

If churches tend to cluster around stations, the *observed* cross K function for the given two sets of points will be larger than the *expected* cross K function to be obtained if churches are uniformly and randomly distributed.

In the case of the K function method, the expected function is obtained from random Monte Carlo simulations, but in the case of the cross K function method, the expected function is analytically obtained. This was shown by Okabe and Yamada (2002).

Figure 10.10 shows the observed cross K function (the black line) and the expected cross K function (the gray line) for the churches and stations in Shibuya-Shinjuku. The black line is above the gray line within 5000 meters, and it is therefore concluded that churches in this area tend to be clustered around transport stations.

10.7 Network Voronoi Diagram

In spatial analysis, there is often a need to estimate the service areas of facilities, such as post offices. Precise estimation is not easy, but a first approximation can be derived from the network Voronoi diagram (Okabe et al.,

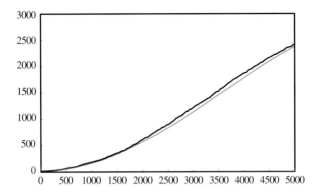

FIGURE 10.10
The observed cross K function (the black line) and the expected cross K function (the gray line) of churches with respect to stations in Shibuya-Shinjuku, Tokyo.

2000). In definition of this, let $q_1, \quad , q_m$ be post offices located on the street network N, and $d(p, q_i)$ be the shortest-path distance from an arbitrary point p on the network N to a post office at q_i. In these terms, we define a sub-network, $V(q_i)$, as:

$$V(q_i) = \{p \mid d(p, q_i) \leq d(p \mid q_j), i \neq j, j = 1, \quad , m\}, i = 1, \ldots m, \qquad (10.4)$$

implying a sub-network in which the nearest post office is q_i. The network Voronoi diagram, V, for the generator points $q_1, \quad , q_m$ is the set of the resulting sub-networks, i.e., $V = \{V(q_1), \quad , V(q_m)\}$. If it is assumed that residents use the post office nearest to their homes, $V(q_i)$ shows the service area of the post office at q_i.

Figure 10.11 shows an example of the network Voronoi diagram constructed using Tool 4. The black circles are the post offices in Shibuya, and the white circles indicate the boundary points between two adjacent Voronoi sub-networks.

10.8 Network Huff Model

One of the most important tasks in retail marketing is to estimate the probability of consumers electing to buy at a particular store selected from among many such stores in a city. The Huff model (1963) provides this choice probability. The model was originally formulated on a plane and later extended to a network by Miller (1994) and Okabe and Okunuki (2001).

To set out the network Huff model explicitly, consumers' houses are considered located at $p_1, \quad , p_n$, and stores lie at $q_1, \quad , q_m$ on a street network.

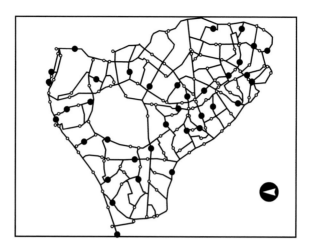

FIGURE 10.11
Network Voronoi diagram of post offices (black circles) in Shibuya, Tokyo. White circles indicate
the boundary points between two adjacent Voronoi sub-networks.

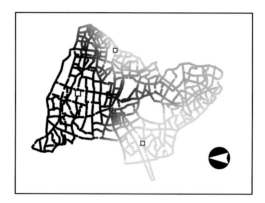

FIGURE 10.12
Probability of the store at the large square being chosen. Small squares mark the other stores.

Let $d(p_i, q_j)$ be the shortest-path distance from a house at p_i to a store at q_j,
and a_j is the magnitude of attractiveness (e.g., the floor area) of the store
at q_j. The probability, $P(p_i, q_j)$, of a consumer at p_i choosing the store at q_j
among the m stores is given by:

$$P(p_i, q_j) = \frac{a_j / d(p_i, q_j)^{\alpha}}{\sum_{k=1}^{m} a_k / d(p_i, q_k)^{\alpha}} \tag{10.5}$$

Tool 13 computes this probability. Figure 10.12 shows an example where
the squares indicate stores, and the density of gray tint indicates the proba-

bility of the store at the large square being chosen. Black indicates a high probability.

10.9 Conclusion

Although, as shown in the introduction, there are numerous network spatial events that attract study by scholars of the humanities and social sciences, spatial analysis of those events began only recently (e.g., Yamada and Thill, 2004; Spooner et al., 2004). One reason for this delay was a lack of tools. SANET now provides a toolbox to assist researchers who wish to analyze spatial events on a network. We hope to hear about successful applications of SANET.

Acknowledgments

We thank T. Ishitomi, K. Okano, and C. Mizuta at Mathematical Programming Co., Ltd. for coding SANET. This development was partly supported by Grant-in-aid for Scientific Research No. 10202201 of the Ministry of Education, Culture, Sports, Science and Technology of Japan.

References

Arapoglou, V., The governance of homelessness in the European South: spatial and institutional contexts of philanthropy in Athens, *Urb. Stud.*, 41, 621–639(19), 2004.

Bandaranaike, S., Graffiti hotspots: physical environment or human dimension, Graffiti and Disorder Conference convened by the Australian Institute of Criminology in conjunction with the Australian Local Government Association, Brisbane, 2003.

Cope, J.G. and Allred, L.J., Illegal parking in handicapped zones: demographic observations and review of the literature, *Rehabil. Psychol.*, 35, 249–257, 1990.

Harries, K., Mapping Crime: Principle and Practice, National Institute of Justice (NIJ) Crime Mapping Research Report, U.S. Department of Justice, NCJ 178919, 1999.

Huff, D. L., A probabilistic analysis of shopping center trade area, *Land Econ.*, 39, 81–90, 1963.

Miller, H.J., Market area delimitation within networks using geographic information systems, *Geogr. Sys.*, 1, 157–173, 1994.

Okabe, A., Boots, B., Sugihara, K., and Chin, S-N., *Spatial Tessellations: Concepts and Applications of Voronoi Diagrams*, 2nd ed., John Wiley, Chichester, 2000.

Okabe, A. and Funamoto, S., An exploratory method for detecting multi-level clumps in the distribution of points — a computational tool, VCM (variable clumping method), *J. Geogr. Sys.*, 2, 111–120, 2000.

Okabe, A. and Okunuki, K., A Computational Method for Estimating the Demand of Retail Stores on a Street Network Using GIS, *Trans. GIS*, 5(3), 209–220, 2001.

Okabe, A., Okunuki, K., and Shiode, S., SANET: A toolbox for spatial analysis on a network, *Geogr. Anal.*, 2005 (to appear).

Okabe, A. and Yamada, I., The *K* function method on a network and its computational implementation, *Geogr. Anal.*, 33, 271–290, 2001.

Painter, K., The impact of street lighting on crime, fear, and pedestrian street use, *Secur. J.*, 5, 116–24, 1994.

Ratcliffe, H.J., Aoristic signatures and the spatio-temporal analysis of high volume crime patterns, *J. Quant. Crim.*, 18, 23–43, 2002.

Ripley, B.D., *Spatial Statistics*, John Wiley & Sons, New York, 1981.

Roach, S.A., *The Theory of Random Clumping*, London, Methuen, 1968.

Shiode, S. and Okabe, A., Network variable clumping method for analyzing point patterns on a network, AAG annual meeting, Philadelphia, 2004.

Snow, J., *On the Mode of Communication of Cholera*, 2nd ed., Churchhill Livingstone, London, 1855.

Spooner, P.G., Lunt, I.D., Okabe, A., and Shiode, S., Spatial analysis of roadside Acacia populations on a road network using the network *K* function. *Land. Ecol.*, 19(5), 491–499, 2004.

Stavric, B., Matula, T.I., Klassen, R., and Downie, R.H., Evaluation of hamburgers and hot dogs for the presence of mutagens, *Food Chem. Toxicol.*, 33, 15–820, 1995.

Tinker, I., *Street Foods: Urban Food and Employment in Developing Countries*, Oxford University Press, New York, 1997.

Yamada, I. and Thill, J-C., Comparison of planar and network *K* functions in traffic accident analysis, *J. Trans. Geogr.*, 12(2), 149–158, 2004.

11

Estimation of Routes and Building Sites Described in Premodern Travel Accounts Through Spatial Reasoning

Yasushi Asami, Takanori Kimura, Masashi Haneda, and Naoko Fukami

CONTENTS

11.1 Introduction

Historical materials, such as documents and excavated evidences, are major sources of historical studies. These sources often show some spatial relationships at the era studied. To efficiently handle such information, geographical information systems (GIS) can be applied.

Sometimes, the location of historical facilities is of particular interest when they are critical to revealing historical evidences. If excavated evidence is

FIGURE 11.1
Isfahan in Iran.

already known, there will be no need to search for the location. There are, however, a number of facilities whose locations are not known. In such cases, historical documents, if any, are the only source of inferring the location. For this purpose, spatial reasoning can be of potential use.

Spatial reasoning is an attempt to infer the spatial location and relationship in spatial-information science. It involves the coding of information concerning spatial locations and inference concerning spatial relationship (Hernandez, 1994; Vieu, 1997). By coding the spatial information, specific spatial relationships can be focused on and topologically or quantitatively analyzed.

This chapter illustrates the applicability of spatial reasoning in historical analysis by showing its application to infer the spatial locations of several facilities written in Jean Chardin's travel account and his walking route in Isfahan in Iran in the 17th century, based on Asami, Kimura, Haneda and Fukami (2002).

Isfahan, an old town located almost at the center of Iran, as shown in Figure 11.1, flourished for about 130 years from 1597, when Abbas I set the town as a capital, until early in the 18th century, when the Safavid dynasty fell. Jean Chardin, a French traveler, visited Isfahan late in 17th century and wrote a travel account based on his experience in the town. His account has been a first class and precious historical source full of detailed and vivid descriptions of the situation in the town (Haneda, 1996).

Despite his detailed description, historians have not decisively located a number of buildings. A notable example is the location of *la Maison de la Douze Tomans* (Twelve Tomans' house), where Chardin lived during his stay in Isfahan. Its location is of particular importance, for it was always the origin of his record. If its precise location can be identified, then locations of other historically important buildings can be inferred with much precision.

Understanding the city structure of Isfahan — including blocks, streets, and locations of religious and urban facilities — in the premodern age is of historical importance, for it leads to the understanding of city planning and urban social structure (Sakamoto, 1980–1981).

11.2 Spatial Data

To start the spatial-reasoning analysis, it is quite important to prepare a base map ready for GIS. It is well-known that the city structure of Isfahan was drastically changed in 1930s as a result of urban-modernization policy in Iran. The city had had no experience of change in large scale before this period. A most reliable map that shows the city structure before the 1930s is one made by Sayyed Reza Khan in 1923 to 1924. An old map, as is often the case, is not very precise in locations of road networks, as well as those of buildings. To make a more accurate map, several facilities in an aerial photograph map, which cannot be moved, are used. The transformed map is digitized to convert to GIS-ready spatial data. With this operation, 337 nodes are recorded. All the roads are expressed by links between nodes, and building use along the roads are input as attribute data of links (Figure 11.2).

The next step for the spatial reasoning is the extraction of descriptions concerning spatial relationships in Isfahan. The travel account used for our analysis is that by Chardin, a French jeweler and traveler (Haneda, 1996). A part of Chardin's travel account describing the route from Mirza Ashraf's residence to "l'Évêché (Bishop's mansion)" is selected for our analysis, for the starting point and the destination of the route, as well as most of buildings appearing in this description cannot be identified by the analysis of the text, and hence the route is not yet identified. Moreover, Chardin seems to have gone forward and then returned to a certain point on the route and walked in another direction. The spatial identification of this route is of great concern in historical studies.

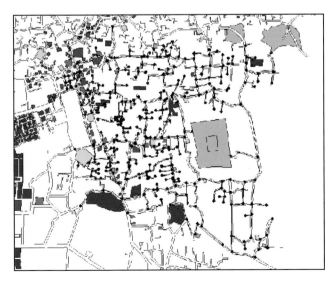

FIGURE 11.2
Digitization of old map (dots in the road network indicates nodes).

There are 51 phrases that describe spatial relationship in this part. The spatial descriptions include relationships such as: (1) specific buildings are neighboring to each other; (2) buildings of particular building use are neighboring to each other; (3) roads branch off into a certain number of roads; (4) turning to left (or right); and (5) going to east (or other direction). These descriptions are related to characteristics of landmarks, crosses, and directions (Haneda, 1996). The building use along the road indicated in the old map is assigned to the road as an attribute. The building name is not shown in the old map. For the details concerning the phrase and judgment of degree of coincidence, refer to Kimura (2002).

11.3 Search for the Route

To derive a route that best fits to the description, two things have to be identified. One is the enumeration of the complete set of potential routes, among which the best route is identified. For this end, a computer program was coded to search all possible routes. It calculates all the possible routes conditional to a set of origins and destinations.

The other important device is the evaluation measure of routes and how they are consistent with the description in the travel accounts. To measure the degree of coincidence of the route to the set of descriptions in the phrases, consistency function, $C(m,n)$, is defined, where m is the identification number of the phrase describing spatial relationship, and n is the order of the link specified. This function takes the value of 0, when the nth link is not consistent with phrase m, while it takes positive value if the link is somehow consistent with the phrase depending on the extent of coincidence. The sum of consistency function over the entire route is defined as degree of fitness with the description of the travel account. Note that even if a route is fixed, there are several ways to match the route to the set of phrases in general. The maximum value is assigned for the degree of fitness among all possible ways of matching.

More specifically, consistency function is defined as follows:

11.3.1 Consistency with Building Use

Since the existence of buildings at the time when Chardin traveled is not known now, consistency is judged based merely on the coincidence of building use along the link. If a building at the age of the Safavids is confirmed, then one more point is assigned than the case that only the building use coincides to the building drawn in Sayyed Reza Khan's map from the 1920s.

11.3.2 Consistency of Branching Off of Road

If the road is branching off into the same or larger number of roads in Sayyed Reza Khan's map, then the link is judged consistent with the phrase. This is because the road network in the 1920s may develop more than that in the age of the Safavids.

11.3.3 Consistency of Turning Direction (Left or Right)

If the link correctly turns to the left or right in accordance with the phrase description, then the link is judged consistent.

11.3.4 Consistency of Direction (North, East, South, and West)

If the corresponding link is closest to the correct direction, then the link is judged consistent.

Consistency function is weighted depending on the reliability by three values, 5, 3, and 1. For example, if the consistency of a building is judged by the coincidence of building use in the 1920s, the reliability of this judgment is rather low, and hence 1 is given for the weight. On the other hand, a road branching off and a description that the road leads to a certain facility are both highly reliable, and 5 is given for the weight. The value of 3 is given as a weight for the case of intermediate reliability (Kimura, 2002).

11.4 Judgment of Neighborhood

Distance in historical documents is described in various forms, from distance measures to ambiguous expressions, such as "near" and "far." A concrete distance measure in Chardin's travel account is the usage of steps, how many steps it is from a certain place to another place. For example, the size of the Royal Square (*Meydan-e Shah*) is described as 440 steps in length and 160 steps in width. A caution should be paid, however, to these numbers. It could be an actual measurement by the author, but it could be also a measurement by impression or rumor. Thus, these numbers have to be examined against the actual size of the plaza.

A vague distance expression appearing in Chardin's travel account is the use of "neighborhood." A place is described in the neighborhood of another place, if Chardin thought it was close enough. This kind of impression can change from situation to situation, but the threshold of usage for maximum distance in the "neighborhood" can be rather stable in a limited context, such as the parts of description in the current analysis.

In the account, *la Maison de la Douze Tomans* is described as follows: "It was located in the neighborhood of the Royal Palace and the Royal Square, and close to British and Dutch East India Company's offices as well as Capucin and Carme mission's houses, and no other place is more convenient." The terms, "neighborhood" and "close" (*à côté de* in French for both), of course indicate that these facilities are near in distance. To estimate up to what distance Chardin used the term, similar expressions were searched for in the travel account.

One description is that "Solimancan's (*Soleyman Khan's*) residence is located 150 steps from the two houses *la Maison de la Douze Tomans* and its next house), and Dutch East India Company's office is next to it." According to this statement, the distance between *la Maison de la Douze Tomans* and Dutch East India Company's office is about 150 steps, but this distance cannot be judged the maximum distance for the range of neighborhood. Since no effective information is available in the description of the travel account, a naïve assumption was made that the maximum distance will be twice as much, i.e., 200 steps, to term "within the neighborhood" in the following.

Chardin reported that the size of the Royal Square was 440 steps in length and 160 steps in width. The Royal Square still exists as it was then, and hence, by measuring the dimension of the plaza, the distance of one step can be estimated. By this procedure, one step is derived as 1.16 meters. By multiplying 300 steps, 347.7 meters can be regarded as the range of neighborhood. Using this distance, the possible area for *la Maison de la Douze Tomans* will be estimated.

11.5 Judgment of Priority in Route Choice

It is natural to assign high priority in the route choice for roads that might have existed at the time of Chardin, such as roads connecting city gates, and the gate and the Royal Palace or the citadel, for more reasonable estimation of the route. To this effect, all the roads are ranked into four levels depending on the possibility of existence in Chardin's age, based on Gaube's estimation (Gaube and Wirth, 1978) and Chardin's description. Moreover, the longer the route on the roads in that age, the more probable, it is judged, that Chardin actually took the route.

Since the degree of fitness is defined as a linear form, as stated in Section 11.3, a longer route tends to gain a higher score than a shorter route. To adjust this tendency, the score for the route that looks longer than expected from the description of the travel account will be diminished.

11.6 Route of Maximum Degree of Fitness and Estimation of Locations of Building

The route that scored the maximum degree of fitness is depicted in Figure 11.3. In the figure, the route consists of two arrows, for Chardin went back

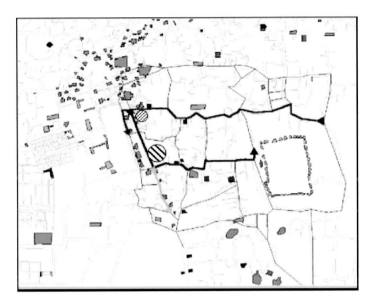

FIGURE 11.3

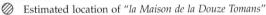 Estimated location of *"la Maison de la Douze Tomans"*

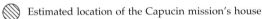 Estimated location of the Capucin mission's house

to a certain point on the route. A route estimated by historians (i.e., Haneda and Fukami, coauthors of this chapter, based on descriptions only without using spatial-reasoning method) is shown in Figure 11.4. A discrepancy is found only in a part to the Royal Square. This is due to the lack of information of buildings in this area, and hence no score can be gained for the historians' estimate route in this area. An examination of historian data in this area will make the estimate of the route more accurate, and the excavation in this area would be of great importance for further historical research.

Location of *la Maison de la Douze Tomans* was estimated based on the phrase that "it was located in the neighborhood of the Royal Palace and the Royal Square, and close to British and Dutch East India Company's offices as well as Capucin and Carme mission's houses, and no other place is more convenient." Using the information in this same phrase, the location of the Capucin mission's house can be estimated, which in turn can be used to estimate the location of the Bazaar with the name of Mahamed Emin (Mohammad Emin), which is recorded to be 30 steps from the Capucin mission's house. A buffering operation for *la Maison de la Douze Tomans* is devised for this process to estimate the location of the Capucin mission's house with the use of the distance for "neighborhood." The estimated locations of *la Maison de la Douze Tomans* are shown in Figures 11.3 and 11.4. Since this house is along the route, where spatial-reasoning results and historians' estimations differ, the location is differently estimated. Figure 11.3 also indicates the location of the Capucin mission's house, which was difficult to estimate by historians without help of spatial reasoning.

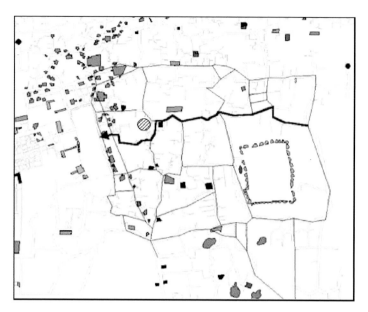

FIGURE 11.4
⊘ Estimated location of *"la Maison de la Douze Tomans"*

11.7 Conclusion

Spatial relationship is seldom a main focus of historical studies, partly due to the lack of analytical tools suited for such studies. Development of GIS, as well as that of methods for spatial analyses, has made it possible to cultivate spatial matters in history. This chapter illustrated an application of spatial reasoning using the GIS technique to estimate the route in Chardin's description of his travels in the 17th century. The resulting route fairly matches historians' estimated route, except for the area close to the Royal Square, where the lack of information for buildings is severe.

Definition of degree of fitness and consistency function would depend on the quality of investigation. Judgment of importance and ranking of roads, for example, can be improved by using information from other various historical sources of the period. In a sense, the spatial-reasoning program can be thought of as a learning process from historical documents. Development of such learning-type evaluation functions can extend this kind of research.

Acknowledgment

Authors of this chapter were benefited from comments by Atsuyuki Okabe, Yukio Sadahiro, and Atsushi Masuyama. This research was supported by Grant-in-aid for Scientific Research by the Ministry of Education, Culture, Science, Sports and Technology of the Japanese government. These are gratefully acknowledged.

References

Asami, Y., Kimura, T., Haneda, M., and Fukami, N., Estimation of Route and Building Sites Described in Pre-modern Travel Accounts Through Spatial Reasoning, papers and proceedings of the Geographic Information Systems Association, 11, 369–372, 2002.

Gaube, H. and Wirth, E., *Der Bazar in Isfahan.*, Wiesbaden, 1978.

Haneda, M., *A Study of Jean Chardin's Description of Isfahan*, University of Tokyo Press, Tokyo, 1996.

Hernandez, D., *Qualitative Representation of Spatial Knowledge*, Springer-Verlag, Berlin, 1996.

Kimura, T., Restoration of Facility Allocation in an Old Map Through Spatial Reasoning: Case Study of Isfahan in the 17th Century, graduation thesis, Department of Urban Engineering, University of Tokyo, 2002.

Sakamoto, T., *Urban Structure and Meydan in the Nineteenth Century Isfahan*,
(I), Shigaku (Historical Studies in Keio University), 50, 367_387,1980;
(II), Shigaku, 51(1_2), 145_158, 1981;
(III), Shigaku, 51(3), 43_79, 1981.

Vieu, L., Spatial representation and reasoning in artificial intelligence, in *Spatial and Temporal Reasoning*, Stock, O., Ed., Kluwer Academic Publishers, Dordrecht, 1997, pp. 5–41.

12

Computer-Simulated Settlements in West Wakasa: Identifying the Ancient Tax Regions — The Go-Ri System

Izumi Niiro

CONTENTS

12.1 Introduction

GIS began to be widely employed in archaeology from the beginning of the 1990s. Early research is represented by *Interpreting Space: GIS and Archaeology* (Allen et al., 1990), the Santa Barbara conference held in 1992 at the University of California (Aldenderfer and Maschner, 1996), and the 1993 Ravello Conference held in Ravello, Italy (Lock and Stancic, 1995). Prior to 1995 "GIS provide[d] archaeologists with a sophisticated means of manipulating spatial data, but offer[ed] limited support for modeling change through time" (Lake 2000). Although it was not directly related to archaeology, the Brookings Institute in the United States developed the Sugarscape model for diachronic research (Epstein and Axtell, 1996). This was a deliberate attempt to simulate chronological change in cultural diffusion, wealth accumulation, and so on using an artificial society comprising agents in 50 × 50 cells. In spatial terms, however, 50 × 50 cells differ greatly from actual societies. Later, M.W. Lake from the University of London used MAGICAL (Multiagent Geographically Informed Computer Analysis) software in an attempt to combine GIS and multiagent simulation based on Mesolithic foraging in Isley, Scotland (Lake, 2000). Attempts to combine spatial analysis and the simulation of chronological change have been gradually advancing.

We have been using GIS in archaeological analyses since around 1993 (Niiro, Kaneda, and Matsushita, 1995). In 2001, we published an introductory volume on GIS archaeology, discussing various examples of spatial analysis in archaeology, mainly using case studies from Japan (Kaneda, Tsumura, and Niiro, 2001). Following this, we have expanded our research to diachronic simulations of demographic change and settlement-formation processes using actual historical spaces.

Various conditions make Japan particularly suited to this type of research. First, since the late 1960s, economic development has been concentrated in a relatively narrow land area, resulting in an astonishing number of rescue excavations. From detailed distribution surveys across Japan, there is a high density of spatial data that link archaeological materials with regional historical reconstructions. Second, data on administrative organization and population statistics remain from the eighth century A.D. onward. There is, for example, a document from 702 that has data on almost all members of some villages with more than 1000 people. Even when viewed on a world scale, this is an important data set of ancient statistics (Farris, 1985). From these ancient demographic data, it has been estimated that the total population of Japan at that time was 5 million. Considering that the detailed population statistics of the Domesday Book in Britain were produced in the 11th century, this represents an important ancient data set. Third, there is the sudden expansion of digital data, including detailed digital-elevation data (DEM: Digital Elevation Model). These three points provide a good context for simulations of actual historical space.

Given these advantages, we have been undertaking an archaeological and historical project in an attempt to develop a computer simulation of diachronic change in ancient society using actual historical spaces. This simulation model is distinctive in that it combines GIS-based spatial analysis and agent-based diachronic analysis. Because the project is ongoing, this chapter focuses on the synchronic part of the current outcome of this simulation. First, we show a simulation model that quantitatively estimates agricultural productivity, considering landforms and historical records. Second, through this GIS-based simulation, we attempt to reveal the intention of the ancient bureaucratic land organization.

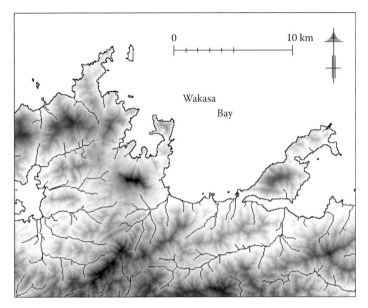

FIGURE 12.1
Geomorphology of the case-study region.

12.2 The Case-Study Region

This chapter uses an area 25 km east-west and 20 km north-south in Wakasa Province, referred to as the Wakasa study region (Figure 12.1). This area forms the western side of Wakasa Bay, which is one of the largest bays on the Japan Sea coast (Figure 12.2). It was chosen for the case study because it is one of the easiest places to compare current geography with ancient bureaucratic land divisions and place names found in the documents. The area is about 100 km north of Nara, the capital at that time, and was known for marine products and salt production. Wakasa, with its good, natural harbors, had long been an important place on maritime transportation routes, and was one focal point for trade with various regions of the Korean Peninsula, as well as being a special area producing marine products for tribute to the court.

In the Wakasa study region, land suited for agricultural production is somewhat limited, the region being known more for fishing and other marine products (Figure 12.1). The southern half of the area is mostly hills, with the highest at 699 m and others more than 400 m dotting the landscape. Settlements are formed along the rivers and on slopes facing the sea, and these rely mainly on rice farming. The northern half, in contrast, has many pen-

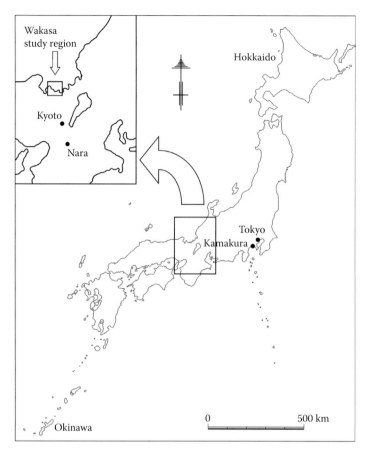

FIGURE 12.2
Location of the Wakasa study region.

insulas, with hardly any land suitable for farming, and dependence on fishing is high.

According to the ancient records, as part of the land division known as the *go-ri* system, six administrative divisions called *go* were established in the Wakasa study region (Tateno, 1995). The *go-ri* system, established in 717, formed the first administrative geographical divisions in ancient Japan and was introduced over quite a short time span until 740. *Go* comprised 50 households, called *ko* (extended families), and were divided into two or three *ri*. In some cases, small units that could not form *ri* were designated as *goko* ("five households"). On average, *go* contained more than 1000 people. The *go-ri* system was established over a wide area of Japan, and the rough location of these divisions can be estimated by linking modern place names with place names recorded in ancient texts and on excavated inscribed wooden tablets (*mokkan*). A long debate has

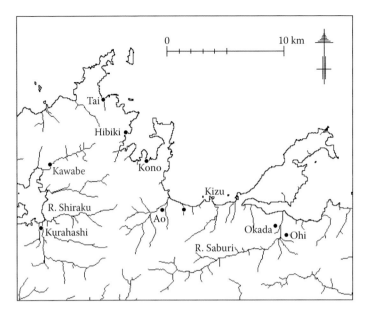

FIGURE 12.3
Modern place names related to ancient *Go-ri* names.

ensued over whether the *go, ri,* and *ko* show the actual nature of villages and families at the time or whether they were predominantly contrived by the authorities. Most scholars agree that administrative influence was strong, but opinions differ over the extent to which the divisions also show the reality of ancient Japan.

The names of the *go* in the Wakasa study region are Shiraku, Kurahashi, Ao, Kizu, Ohi, and Sabu (Figure 12.3). Of these, Ao *go* seems to have the most complex structure. From the texts it is known that Ao *go* was made up of the following *ri* and *goko*: Ao *ri*, Ono *ri*, Kawabe *ri*, Hibiki *goko*, and Tayui *goko* (Tateno, 1995). Of these, the pronunciation of Ono is close to the modern place name of Kono, and Tayui to Tai. Furthermore, the present locations of most of the place names of Ao *go* are known. Looking at the distribution of those names, it is difficult to regard the whole of Ao *go* as having a geographical unity. In particular, Kawabe belongs to a completely different river system and is now in a different prefecture. Kono, Hibiki, and Tai are fishing villages, and it is difficult to see them as being especially closely related to Ao *ri*. From these points, therefore, the composition of Ao *go* has several unnatural aspects. If we could understand the causes of this structural complexity, it should be possible to deepen our understanding of the existing debate over whether or not *go, ri,* and *ko* represent actual ancient villages and families. We would like to investigate this problem using a GIS-based reconstruction of agricultural productivity.

12.3 The Reconstruction of Agricultural Productivity and the Extent of *GO*

With the exception of Hokkaido and Okinawa (Figure 12.2), the representative agricultural system of Japan was a combination of wet rice paddies and the cultivation of vegetables and other crops in dry fields. Domesticated animals played almost no role in this system, except for cattle and horses used in field cultivation. Large-scale cultivation of wheat and barley also was rare. Furthermore, a great deal of energy was expended in obtaining level surfaces for paddy fields. Agricultural colonization progressed with the construction of paddies on plains and in stepped rice paddies on hills. As a result, the agricultural landscape was very different from that seen in Europe and elsewhere. Fields were concentrated along rivers and plains, some hill slopes also being used for dry fields.

Agricultural productivity is related to geomorphology, water, sunlight, soils, and various other complex factors. However, in cases where the environment does not differ greatly, the effect of slope angle is dominant. Using the 50 m grid-elevation data produced by the Japanese Geographical Survey Institute, "slope values" from 0 to 9 were assigned to each grid cell (the accuracy of elevation used in this research was 0.1 m). To be explicit, the slope of each cell was defined as the largest slope between the central cell and the eight cells surrounding it (Figure 12.4a). A slope of 0–2 percent was given the value of 9 (Figure 12.4b). A value of 1 was then subtracted for each subsequent 1 percent increase in slope angle. Slopes of 10 percent or more were valued at 0. Ten percent represents a slope angle of more than 5°. Today, slopes of up to 8° are regarded as possible for rice-paddy construction, but considering the limitations of ancient technology, it was decided to set the lower level of 5° for this research. Regarding elevation, values at 400 m above sea level were assigned 0. Even today, there are hardly any villages at elevations of more than 400 m in this study region. Because cells with an elevation of 2 m or less are potentially subject to high waves and other damage, they were assumed to have a low suitability for agriculture and were valued at 0.

Based on the hypothetical agricultural-productivity data derived in this way, the values of cells that are included in the area centered at each cell with the radius of five cells (about 250 m, as in Figure 12.4c) are summed up. The centers of the solid circles in Figure 12.5 show the cells whose resulting values are the first to the 50th largest, and their diameters correspond to their values. Note that once the *i*th largest cell is determined, the values of the cells within the radius of five cells are set at 0, and so the solid circles do not overlap each other ($i = 1, …, 50$). The gray color in Figure 12.5 shows the degree of slope; the darker the color, the steeper the slope. The actual locations of settlements usually avoid areas suitable for rice paddies and are thus found on the edges of the plains, but initially we want to use

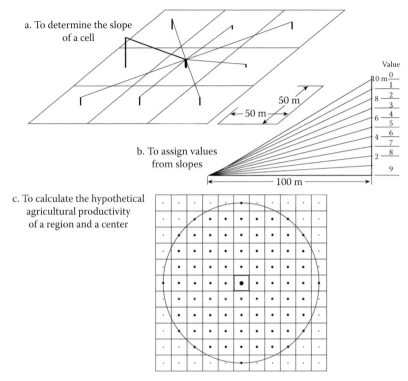

FIGURE 12.4
Method of reconstructing agricultural productivity.

this simulation to search for where settlements were. This distribution appears to some extent similar to the modern distribution of farming villages. The fact that hardly any settlements were drawn in the northern half of this region does not contradict the fact that fishing villages are concentrated here today.

To estimate the area of each *go* and *ri*, we found geographically cohesive areas by examining the agricultural productivity shown in Figure 12.6. Such areas are indicated by circles, the number in each circle showing the value of total agricultural productivity of this region.

According to the documents, it is almost certain that there were two *gos*, Kurahashi and Shiraku, in the area at the left of Figure 12.6. From the total value in this area, which was 11,186, we presume that a *go* has a value of about 5500 (i.e., 11,186 divided by two).

The total for the area of the River Saburi at the bottom right of the map was 10,681. Divided by two, this is 5340, a number close to the value for a single *go* of 5500. According to the texts, Sabu *go* was located on the upper reaches and Ohi *go* on the lower reaches of the River Saburi, which suggests that the values are appropriate. The geomorphology of the Saburi basin shows it was constricted, and the river can be divided into upper and lower

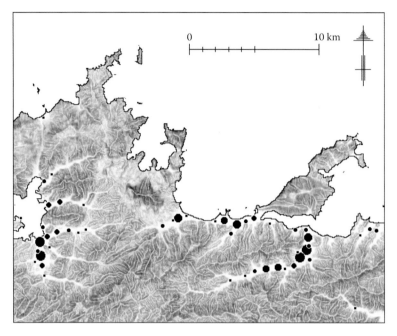

FIGURE 12.5
Settlement distribution from the simulation.

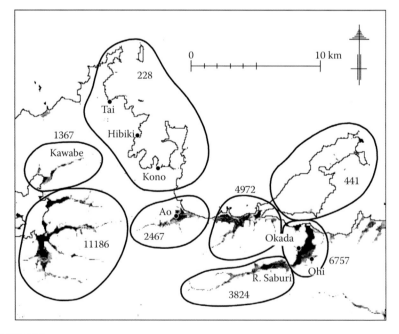

FIGURE 12.6
Estimated divisions based on productivity.

reaches. The values of these two areas are very unbalanced, the upper basin having a value of 3824 and the lower 6757. According to the texts, Okada *ri*, which was located on the lower left bank, was part of the upper basin Sabu *go*. At the same time, the Ohi Shrine, which is believed to have belonged to the lower Ohi *go*, was located on the opposite bank from Okada and slightly upstream. The upper reaches of the River Saburi were probably difficult to combine with a geographically separate watercourse, and therefore the shortfall in households in the upper basin was made up with Okada *ri*. From this, it can be assumed that the number of households had priority over landforms in the division of *go*. Kizu *go* has a value of 4972, which seems a little low, but fishing plays a role in this area today, and this is probably a normal value.

Ao *go*, at the center of Figure 12.6, is even more complicated. Ao *ri* has a value of 2467, which is only about half of that of a *go*. The fishing villages of Kono, Hibiki, and Tai were added, but this was still not enough, and the area of the separate watercourse of Kawabe was probably required to form a single *go*. The combined value was 4062, which is about 75–80 percent of a standard *go*. The remaining 20 percent or so was probably made up by households engaged in other activities, including fishing.

12.4 Results and Discussion of the Case Study

From the above analysis, we have concluded that the area of a *go* was established according to the following processes:

1. Officials located regions that were cohesive with respect to natural geomorphology and counted the number of households (recall Figure 12.6). The upper River Saburi region is such a region.

2. If the number of households was not sufficient to form a *go*, neighboring regions were added without due attention to their close relationships, for example, the addition of Okada from the lower Saburi to the upper river, and the addition of areas from a different watercourse to Ao *ri* and Ao *go*.

It has already been mentioned that a *go* was made up of 50 households, and each of those households comprised an average of more than 20 people. This is generally referred to as an extended household, but there is a sharp difference of opinion over whether this size reflects the actual nature of families at that time or whether it was artificially created by the bureaucracy for taxation or other purposes. However, according to the results of a simulation of demographic change based on ancient family registers, conducted by a member of our project, Katsunori Imazu, at a time of high fertility and high mortality, when average life expectancy did not reach 30, a large family would have been required in order to function adequately as a production unit (Imazu, 2003). Therefore, it is difficult to believe that actual households were registered without modification. It can be assumed that, first, actual households that consisted of more than 20 people were registered in a *ko*; second, if the size of a household was fewer than 20 people, close kin were added, or persons were added according to mechanical administrative rules. This method was similar to that employed for constructing *go*.

This conclusion had been suggested to some extent by previous research, but it has usually been debated based only on inaccurate intuition. In the present study, relatively simple methods were used to derive estimates that provide a quantitative basis for the discussion. This is one significant aspect of this study. However, this type of quantitative discussion would be practically impossible without the application of GIS, and we consider it highly significant that this is the first study that shows the superiority of GIS over existing approaches in a concrete and persuasive way.

Applying the method used in this case study, it should be possible to estimate the location of *go* and settlements whose precise positions are not recorded in the texts. Where the exact locations of *go* and *ri* are known, comparisons with hypothetical agricultural productivity estimated from landforms will enable further understanding of ironworking and other crafts in addition to the fishing discussed here, and also will enable estimates of the roles of other economic activities.

Clarification of regional population capacity based on agricultural productivity as discussed above makes it possible to link these synchronic data with simulations of diachronic change. As noted, Imazu's simulation of ancient demographic change used agent-based analysis to reconstruct quite detailed models of births, deaths, and marriages. Based on the population capacity used here, if we apply Imazu's results using the Brookings Institute Sugarscape model with 50×50 cells as mentioned in the introduction, it should be possible to develop spatial simulations for particular regions.

12.5 Broader-Scale Analysis

The analysis described above is also applicable to regions that are larger than the Wakasa study region. To identify possibilities for future research, we would next like to attempt the reconstruction of agricultural productivity over broader areas of Japan and of hypothetical regional centers (Figure 12.7). Japan's capital city has moved from west to east, from Nara and Kyoto to Kamakura and Tokyo (Figure 12.2). To clarify the movement of these centers and the regional structure of ancient Japan, we would like to estimate agricultural productivity from landforms and compare the regional centers.

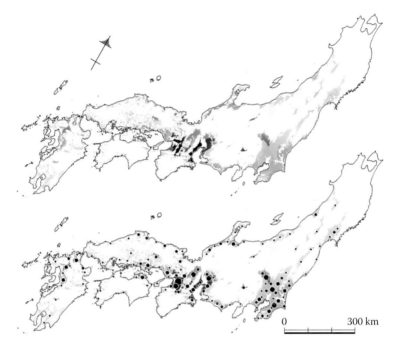

FIGURE 12.7
Hypothetical agricultural production and regional centers in the Japanese Archipelago.

When analyzing larger regions, it is necessary to add slightly different methods to those used for Wakasa. Most important is the difference in temperature from south to north. Assuming that average temperature declines with increasing latitude, and that temperature drops by 0.6°C with each 100 m of elevation, the values assigned from geomorphology have to be adjusted to temperature calculated solely from the latitude. In the Japanese archipelago, ocean currents influence the temperature, but this was not considered. Our first map of estimated agricultural productivity, which was based on slope and temperature, was significantly different from our image of production based on the texts. This appears to be because of geographical conditions, such as soils and different histories of regional development. We therefore adjusted it, using as a guide the text on the size of the area under agriculture of each ancient province (*kuni*) written in the 10th century. The estimates of agricultural productivity obtained in this way were close to the picture known from documents and other data (Figure 12.7: upper).

Based on these estimates of agricultural productivity, regional centers were plotted using sums of values for agricultural productivity, as with the Wakasa study (Figure 12.5). This map (Figure 12.7: lower) provides a good indication of the regional power balance in ancient Japan. Using these results, it will be possible to conduct further simulations of interregional exchange and the circulation of goods.

12.6 Conclusion

In this chapter we have shown a simulation model that quantitatively estimated agricultural productivity, taking account of landforms and historical records, using GIS. Through this GIS-based simulation, we have revealed the purpose of the ancient bureaucratic land organization. Of course, the configuration of actual societies was not completely determined by landforms. However, the effects of factors (including cognitive aspects) other than environmental (including landforms) can be revealed only through a thorough analysis of characteristics determined by the environment. The construction of agent-based simulations that fully consider environmental factors is a task for the future.

References

Aldenderfer, M. and Maschner, D.G.. Eds., *Anthropology, Space, and Geographic Information Systems*, Oxford University Press, New York, 1996.

Allen, K.M.S., Green, S.W., and Zubrow, E.B.W., Eds., *Interpreting Space: GIS and Archaeology*, Taylor & Francis, London, 1990.

Epstein, J.M. and Axtell, R., *Growing Artificial Societies: Social Science from the Bottom Up*, MIT Press, Cambridge, 1996.

Farris, W.W., *Population, Disease, and Land in Early Japan, 645–900*, Harvard University Press, Cambridge, Massachusetts, and London, 1995.

Imazu, K., Villages and regional society in ancient Japan, *Quarter. Arch. Stud.*, 50(3) pp 57–74, 2003.

Kaneda, A., Tsumura, H., and Niiro, I., *Geographical Information System for Archaeology*, Kokon Syoin, Tokyo, 2001.

Lock, G. and Stancic, Z., Eds., *Archaeology and Geographical Information Systems: A European Perspective*, Taylor & Francis, London and Bristol, 1995.

Lake, M.W., MAGICAL computer simulation of Mesolithic foraging, in *Dynamics in Human and Primate Societies: Agent-Based Modeling of Social and Spatial Processes*, Koler, T.A. and Gumerman, G.J., Eds., Oxford University Press, New York and Oxford, 2000.

Niiro, I., Kaneda, A., and Matsushita, O., (1995) Potential application of GIS to Japanese archaeology, *Quarter. Arch. Stud.*, 42(3) pp. 92–99, 1995.

Tateno, K. Reconstructing study on the Gori system: with principal reference to the Nijo Oji wooden writing tablets, in *Advances in the Study of Cultural Properties*, Dohosya, Kyoto, 1995.

13

Site-Catchment Analysis of Prehistoric Settlements by Reconstructing Paleoenvironments with GIS

Hiro'omi Tsumura

CONTENTS

13.1 A Brief Review of Spatial Archaeology

In archaeology, methods for reconstructing prehistoric settlements have been discussed from various viewpoints. Until the 1970s, "Marxist or Darwinist" archaeologists (Dark, 1995) commonly emphasized the importance of the socio-economic contexts of prehistoric societies. However, after the 1980s, many Western archaeologists questioned these approaches. As alternatives, "new" or "process" archaeology (Dark, 1995) emerged, emphasizing a "human–nature interaction"; that is to say, settlements were considered to have originated through the interaction between human behavior and the environment.

Along with this stream of thought, Hodder and Orton (1976), Clarke (1977), and others developed spatial archaeology, which emphasized the importance of spatial attributes in archaeological information. In the early stages of its development, archaeologists encountered difficulty in treating such spatial attributes. In the 1990s, however, this problem was significantly overcome through the introduction of geographical information systems (GIS), which were developed together with rapid progress in the refinement of personal computers and spatial-information science (e.g., Allen, Green, and Zubrow, 1990; Lock and Stancic, 1995; Kaneda, Tsumura, and Niiro, 2001).

Since GIS enabled the systematic integration of archaeological, geographical, and environmental information, those archaeologists who were interested in prehistoric settlements began applying GIS to field research and theoretical studies. For example, Gaffney and Stančič (1991) chose the island of Hvar in Croatia as a pilot-study area, and carried out comparative studies concerning the "human–nature interaction" upon Roman settlements. Their approach took joint account of human material culture and natural landscape factors, and they disentangled the complicated mechanisms of "human–natural interactions" by using GIS.

In the 2000s, GIS archaeologists began to develop several new perspectives. For instance, Spikins (2000), Tsumura (2001), Indruszewski (2002), Ceccarelli and Niccolucci (2003), Clevis et al. (2004), and others considered that paleolandscape reconstruction was indispensable in order to understand the dynamics of human ecology. Kamermans (2000), Verhagen and Berger (2001), Doortje (2003), and others proposed predictive modeling and simulation of the paleoenvironment, expecting that it would overcome the limitations of inductive approaches. Crescioli, D'Andrea and Niccolucci (2000), Hatzinikolaou et al. (2003), and others attempted an integrated approach using fuzzy logic that may clarify errors within a deterministic interpretation. Reynoso and Castro (2004), Reynoso and Jezierski (2002), and others adopted the chaos theory and simulation methods in an attempt to establish a deductive method for explaining prehistoric phenomena.

In this chapter, we illustrate GIS-based methods for reconstructing the paleoenvironment that has been developed in the Sannai-Maruyama site archaeological project. This project was the first large-scale project carried out in Japan with the aim of understanding the "human–nature interaction" concept history. The chapter consists of seven sections, starting with this review. Section 13.2 outlines the Sannai-Maruyama project, and Section 13.3 describes the modern environment of the area surrounding the Sannai-Maruyama site. Section 13.4 discusses the construction of the spatio-temporal database used for the project. Section 13.5 attempts, through analogical inference from geological data and the present ecological environment, to reconstruct the paleoenvironment of the Sannai-Maruyama settlement, which lasted for 1700 years about 5000 years ago. Based on this reconstruction, Section 13.6 attempts to determine the catchment area of the Sannai-Maruyama settlement and looks at the distinctive characteristics of the Sannai-Maruyama site when it is compared with a similar site. The last section summarizes the analytical methods and the major results derived from the project.

COLOR FIGURE 2.11
Creating various routes by using spatial-hyperlinks.

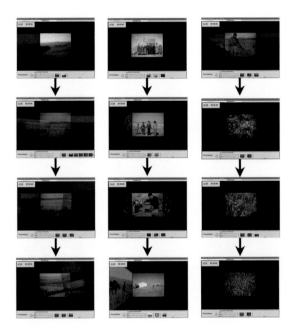

COLOR FIGURE 2.15
Hyperphoto network created from photos taken in Mongolia — spatial relationship (left),
temporal relationship (center), and semantic relationship (right).

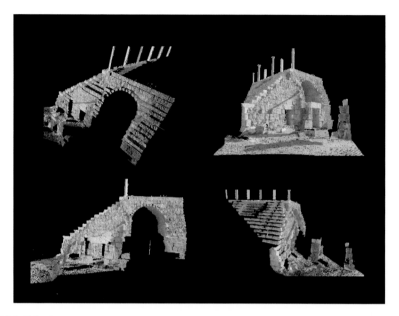

COLOR FIGURE 3.16
Perspective images of the 3D + textured data of stepped stadium from different viewpoints.

COLOR FIGURE 14.10 (a and b
Typical buildings in the survey block (left: Bigam/right: zinim).

COLOR FIGURE 18.8
Bird's-eye view of *Edo* city (1843).

COLOR FIGURE 18.9
Bird's-eye view of Tokyo in the *Meiji* period (1888).

Nihonbashi

COLOR FIGURE 18.10
Reproduction of the landscape depicted by Hiroshige's *Ukiyo-e* (a) *Nihonbashi-Yukibare* (Source: The Money Museum of UFJ Bank), (b) reproduction of the landscape, (c) *Tenpou-Edo-Zu* map (1843), and (d) current scene.

COLOR FIGURE 18.11
Reproduction of a landscape not depicted by *Ukiyo-e* (1).

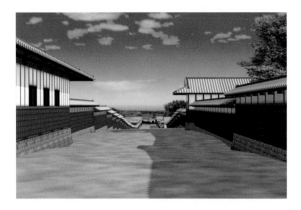

COLOR FIGURE 18.12
Reproduction of a landscape not depicted by *Ukiyo-e* (2).

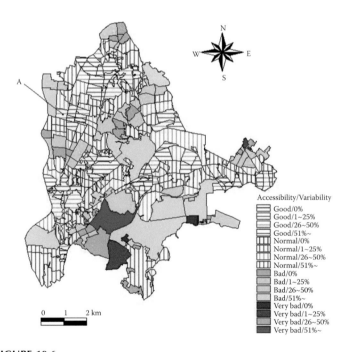

COLOR FIGURE 19.6
Bivariate map of accessibility at district level and its internal variability.

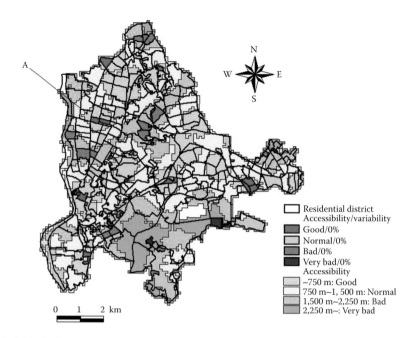

COLOR FIGURE 19.8
Accessibility map at variable spatial level.

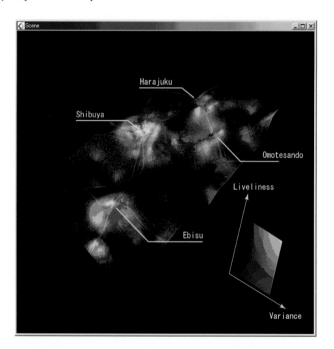

COLOR FIGURE 20.2a
The image of Shibuya: a) liveliness and b) elegance.

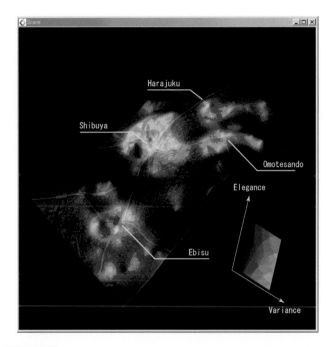

COLOR FIGURE 20.2B

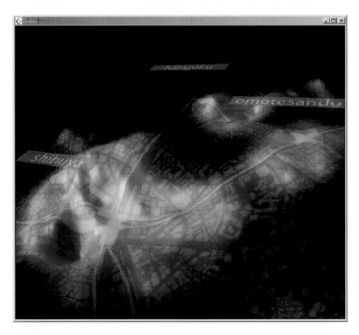

COLOR FIGURE 20.3
The image of Shibuya.

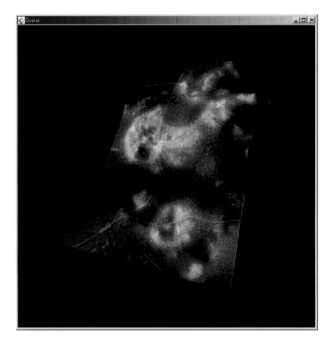

COLOR FIGURE 20.4a
Elegance of Shibuya in the a) daytime and b) nighttime.

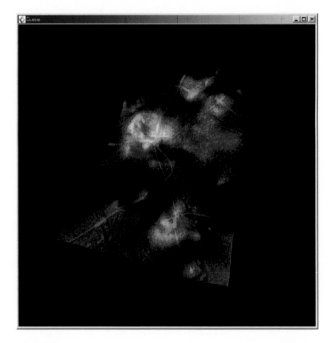

COLOR FIGURE 20.4b

13.2 The Sannai-Maruyama Archaeological Project

The Sannai-Maruyama site, which was designated as a special national historical site in November 2000, was settled from the early to late stages of the Jomon period, a Japanese Mesolithic culture. Many archaeological artifacts, including a great quantity of Jomon pottery, stone lithic articles and ornaments, clay figurines, pit dwellings, storage chambers, clay-mining pits, and graves, were unearthed during excavations carried out since 1992. Radiocarbon dating showed that people lived at this site for 1700 years between 5900 to 4200 B.P. The nature of the settlement that can be discerned following excavation at this site is different in two respects from that which archaeologists imagined before excavation. First, the number of dwellings in a typical prehistoric village is considerably larger than we imagined. We had considered that 5–10 houses, with 20–50 inhabitants would be normal. However, 50 to 100 houses were discovered to represent one archaeological phase, suggesting that 200–400 people lived together. Second, the life span of the villages in the site was much longer than we originally thought. Most were generally maintained for one to three generations, or 50–100 years. People lived at the Sannai-Maruyama site for 1700 years continuously. Maintaining such a large-scale settlement must have been difficult for a Mesolithic society whose survival basis was subsistence hunting and gathering.

The following questions arise:

What kind of subsistence strategies supported the size and duration of the Sannai-Maruyama site?

What kind of environmental factors allowed the people to live with such a lifestyle?

To answer these questions, interdisciplinary studies were undertaken in collaboration with researchers in ecology, geology, geography, zoology, biology, and, of course, archaeology. GIS played a key role in integrating these many types of data.

13.3 Present Nature of the Area Surrounding the Sannai-Maruyama Archaeological Site

To appreciate the location and environmental characteristics of the Sannai-Maruyama archaeological site and the surrounding area where the prehistoric Sannai-Maruyama people lived, the current environment is examined.

The study area (the rectangular area in Figure 13.1b) is in the northern part of Honshu, the main island in Japan (Figure 13.1a). The area measures

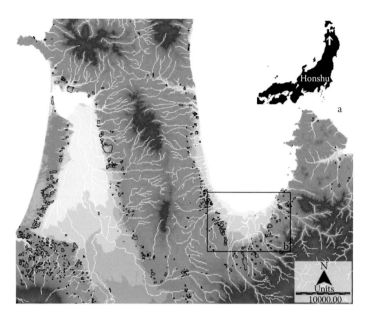

FIGURE 13.1
The archaeological site distribution map on DEM.

about 2880 square kilometres, and contains some 700 archaeological sites from the Jomon period (see Figures 13.1 and 13.2). The Sannai-Maruyama site is located on what is now diluvial upland, with an altitude of 10–20 m. This upland is formed upon a bedrock of pyroclastic material typified by volcanic ash, which has been eroded into an undulating topography. This upland and the Aomori Plain are separated by the Nyunai Fault (Figure 13.2). The Okidate River (Figure 13.2) flows from the border of the Sannai-Maruyama site into the Aomori Bay. The geomorphic features of the east of the Aomori Plain differ from the west because the Komagome and Nonai rivers (Figure 13.2), which flow through the piedmont alluvial plain of Hakkouda, provide a large quantity of volcanically derived sediment to the Aomori Plain, which has created several alluvial fans on the east.

The vegetation of the uplands and hills surrounding the Aomori Plain consists of oak (*Quercus crispula*) or planted forest of evergreen conifers. This oak is thought to be second-growth forest following a massive deforestation of Japanese beech (*Fagus crenata*).

In addition to Sannai-Maruyama, we examined one further locality (which will be referred to in Section 13.6). This comparison widens our perspective and clarifies some distinctive characteristics of the Sannai-Maruyama site.

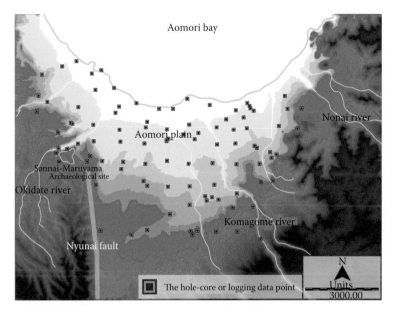

FIGURE 13.2
The hole-core distribution map and geomorphic features on the Aomori plain.

13.4 Construction of a Spatio-Temporal GIS Database

The archaeological database consists of many data sets. The first such set includes the geographical, geological, and geomorphological data of the Sannai-Maruyama Mesolithic site and its surroundings. It includes rivers (lines), soils (polygons), and climatic zones (polygons). For comparisons, the first data set includes observations on other sites and surroundings. The first data set is the base map for the second data set.

The second data set consists of archaeological survey information, including the boundaries of sites (polygons) and the spatial distribution of pottery, stone artifacts, dwellings, and other remains. These data were recorded with their locations, but we often met the following difficulties.

The boundary detail was obtained by digitizing site areas recorded on paper maps. Considerable difficulty arises when the boundaries are superimposed on the GIS geographical-coordinate system, because almost all paper maps of archaeological sites employ arbitrary coordinates. Old maps of sites were recorded on handwritten memos or they used local surveys that were independent of the standard geographical-coordinate system. Although control points were recorded, these did not relate to formal reference points. The location data or addresses of sites were not always recorded because many of the paper maps were without textual data. Moreover, plural

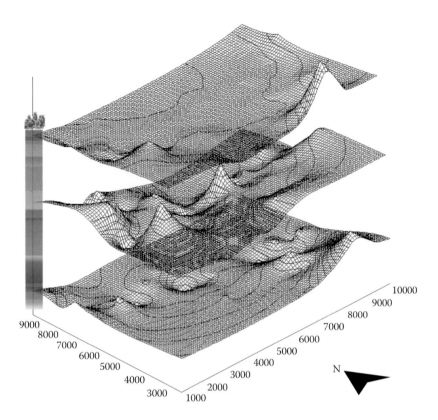

FIGURE 13.3
The 3D geological-layers model of the Aomori plain.

sites often have the same location, and an address in rural areas tends to indicate such a large area that it is difficult to identify the precise locality. It is therefore not a straightforward matter to superimpose the site data recorded in old paper maps and memos onto the precise geographical-coordinate system of GIS.

Another difficulty arises from the manner in which the site area was divided when it was occupied. Since all the sites are not fully excavated, the village area of that time is not clearly identified. Even if the upper sedimentary layers were removed to expose an ancient surface, it is difficult to now see the boundaries of the site area, because precisely discernible boundaries did not exist in primeval societies. In order to construct digital base maps, we therefore determined a probable site area that was consensually estimated by many researchers, thus taking the second-best method. This approach is very simple and unsophisticated, but it surely converts raw archaeological information into digital form.

The third data set consists of the attribute data of the features in the first and second sets. For example, data are the duration of habitation in the

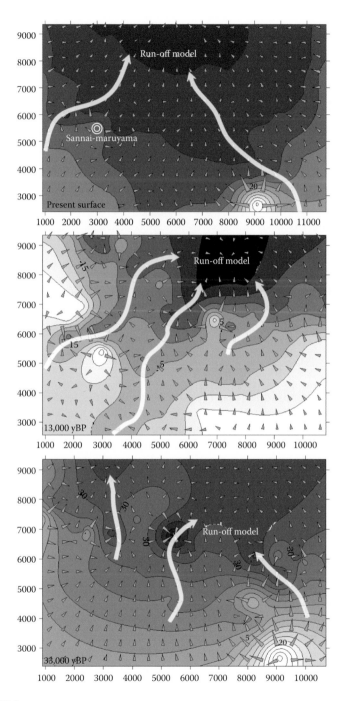

FIGURE 13.4
The DEMs of the paleoground surface under the Aomori plain.

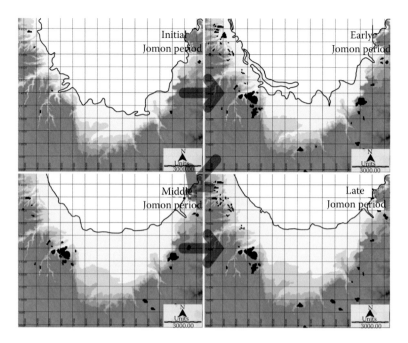

FIGURE 13.5
A simulation model of the sea-level fluctuations.

villages and the numbers and types of pottery, lithics, dwellings, and other archaeological remains. Note that the attributes include time data.

These three data sets are combined through the IDs of features, and so we can easily reference the location, archaeological time, and attributes of every aspect. This system provides a spatio-temporal database for archaeological studies.

13.5 Reconstruction of the Paleosynecology

Archaeologists used to infer the manner of subsistence within a site and the environment around it just from its waste. However, in order to understand locational characteristics of a site, we should also examine its paleosynecological background. From the excavation at Sannai-Maruyama, the waste shows what natural resources the Sannai-Maruyama people used over 1700 years. The most prominent discarded materials are marine, such as shell or coastal-fish remains, as well as the remnants of terrestrial resources, such as chestnut or cervid husks.

Mesolithic people knew where to find sustenance without maps. They recognized and knew the requisite natural conditions. For example, they understood that chestnut tends to grow in sunny and well-drained areas, and that the male deer prefers to stay alone on steep slopes — a characteristic that makes him easy to hunt. By overlaying these ecological and, in particular, the geoecological characteristics of resources, we can reconstruct a cognitive map from them.

We carried out a synecological simulation of chestnut and cervid growth, since these were main resources in the Jomon subsistence, by using GIS. We attempted to estimate the amounts consumed using biological data, but this was difficult, because the natural environment also changed, following along with human evolution. We therefore treated these synecological factors quantitatively in the following manner.

To simulate the distribution of terrestrial resources on the palaeo DEMs (which was reconstructed in Section 13.4), we considered 12 growing conditions for chestnuts and eight for cervids in relation to geomorphological features. For example, chestnut is the favored crop in an area that is not damp, has less than 30 degrees of slope, faces southeast, and is well-drained. Cervids like an area with an undergrowth of bamboo grass, with less snowfall in the woodland. Each condition was recorded on the palaeo DEMs in terms of 1 or 0, implying that the condition is satisfied or not. The resulting 12 layers for chestnuts and eight layers for cervids were overlaid using GIS.

As shown above, our method explicitly considers the uneven distributions of resources. Although this approach tends to be qualitative rather than quantitative, it is more valid than the traditional method that assumes inexhaustible and ubiquitous resources. The locational aspects were ignored.

Figures 13.6 and 13.7 show the probability of chestnut- and cervid-growth distribution. Each legend indicates the cumulative number of Boolean layers, implying the number of satisfied conditions. We notice from these figures that the cumulative numbers for chestnut decrease within mountainous areas, increase along rivers, and are generally high around the areas of numerous Mesolithic sites. Conversely, cervids conspicuously occur within mountainous areas, such as the foothills of Mount Iwaki, and are less prevalent in areas where villages existed.

These results have significant implications when considering economic specialization in the subsistence economy of Mesolithic societies, because we understand that Mesolithic nonagricultural, horticultural subsistence was more strongly influenced by environmental conditions than Neolithic agricultural subsistence.

13.6 Site-Catchment Analysis of the Reconstructed Paleoenvironment

Traditionally, the function of a site is determined from remains found in a site. For example, if a shell mound at a site was large and the evidence of residence was small, the site was inferred to be a shellfish-gathering place.

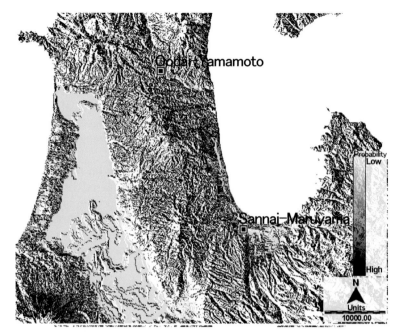

FIGURE 13.6
A distribution map of chestnut-growth probabilities in the target area.

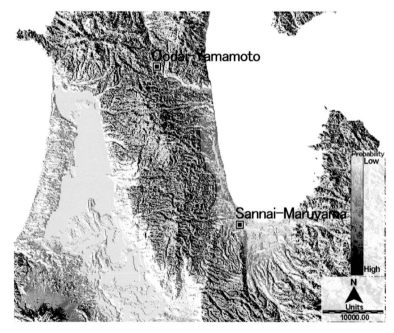

FIGURE 13.7
A distribution map of cervids-hunting probabilities in the target area.

If the size of a site was small and hunting gear, such as arrowheads, were found, the site was inferred to be a hunting base. However, these inferences only considered the physical remains of the hunt.

We should deepen the historical understanding of a site by interpreting the physical remains in a social context. For example, even if the shell mound of a site was large and remains of residence were small, the site might not have been solely a shellfish-harvest place, but it could also have been an ordinary settlement if there were no other such settlements around the site. We should try to understand the function of a site considering the social and spatial relations among sites.

To consider the function of a site, we employed site-catchment analysis, which is one of the most important analyses in archaeology. The key concept used in this analysis is the site-catchment area. This means the area in which the economic activities of everyday life are conducted. In other words, the catchment area implies the district surrounding a site from which the resources for everyday life are obtained. This area should be distinct from political areas and sociological territories. If we can estimate the quantities, as well as qualities, of a catchment area, we may understand the historical context of a site, as well as that of the remains found in the site, more correctly.

In traditional archaeology, researchers usually estimate the site-catchment area by drawing circles with radii of several walking distances. This method implicitly assumes that a plain has no obstacles, such as mountains, valleys, or rivers, such that the same physically distant points in any direction would have the same walking time. In reality, this is not so, as the distance walked in a certain time varies according to geomorphological constraints. Consequently, the form of the site-catchment area becomes an irregular circle or oval. Fortunately, GIS can easily consider these factors, because GIS contains the tools for measuring the angles of slopes, distances and the necessary effort. Remarking upon this advantage of GIS, Renfrew and Bahn (2000) state that site-catchment analysis with GIS is very practical, because the extent of environmental resources within a one- or two-hour walk can be easily estimated. It is no exaggeration to say that all the historical interpretations of an archaeological site are based on the results of site-catchment analysis that is prerequisite in considering the social and cultural functions of a site.

To clarify the particular characteristics of Sannai-Maruyama, we compared it with the Oodai-Yamamoto site. This is located further inland, it is smaller, and its occupation period is shorter than that of Sannai-Maruyama. We nevertheless consider Oodai-Yamamoto comparable in that people lived there continuously since the upper Palaeolithic period.

We simulated the site-catchment areas of both locations using the methods outlined in Sections 13.4 and 13.5 (DEM and slope angles). Figure 13.8 shows the site-catchment area of Sannai-Maruyama, and Figure 13.9 shows that of Oodai-Yamamoto. The irregular circles reflect the limits to which walkers could reach at the same cost in effort (note that the number, say, 9, of walking cost in Figure 13.9 means the equivalent of 9 km of walking on a flat land-

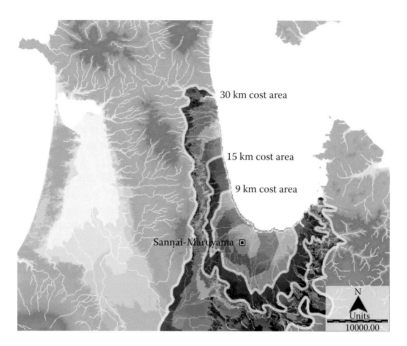

FIGURE 13.8
The site-catchment area of the Sannai-Maruyama site.

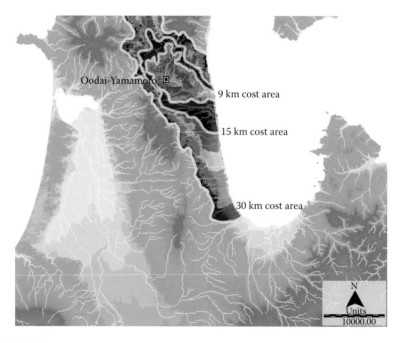

FIGURE 13.9
The site-catchment area of the Oodai-Yamamoto site.

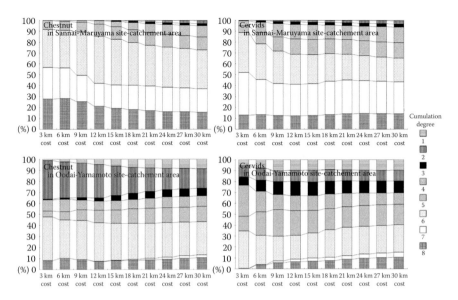

FIGURE 13.10
Each component ratio of the acquisition probability about chestnut and cervids.

scape). We notice from these figures that the site-catchment area of Oodai-Yamamoto is smaller than that of Sannai-Maruyama, because the former is more mountainous than the latter, which is located near the Aomori Plain.

We next overlaid these site-catchment areas on their paleosynecological environments obtained from the simulation in Section 13.5. Figure 13.10 shows the probabilities of acquiring chestnut and cervids in the site-catchment areas of Sannai-Maruyama (upper left in Figure 13.10) and Oodai-Yamamoto (lower left), respectively, with regard to their walking cost. We first notice from this figure that the efficiency of acquiring chestnut per walking cost in the site-catchment area of Sannai-Maruyama is higher than that at Oodai-Yamamoto. Second, we notice that the efficiency of acquiring cervids in the site-catchment area of Sannai-Maruyama (lower left in Figure 13.10) reduces faster than that at Oodai-Yamamoto (lower right) with respect to walking cost. These findings imply that intensive horticulture is more appropriate to the area surrounding Sannai-Maruyama, whereas diversified subsistence is more suitable for the area of Oodai-Yamamoto.

From the above considerations, we infer that people in Sannai-Maruyama village specialized in intensive horticulture by harvesting chestnuts and other vegetable resources in the surrounding district and, at the same time, they developed a social network of intervillage trade. Such a large and long-maintained settlement would not have been possible through primitive self-sustaining subsistence alone. One of the most notable findings is that even in an area of intensive horticulture, there existed settlements whose size was as large as those of Neolithic societies. It should be noted, therefore, that large settlements do not always imply Neolithic settlements.

13.7 Beyond Spatial Archaeology

This chapter presents a new archaeology, which is spatial archaeology enhanced by GIS, or otherwise GIS-based spatial archaeology, which overcomes traditional limitations. Having heard "GIS-based," one might consider that the new spatial archaeology is merely an ordinary application of GIS, but it is more than that. The new spatial archaeology integrates many types of information: on geomorphology, behavioral science, synecology, geology, and other related disciplines. In this integration, GIS plays a crucial, pivotal role.

Another important aspect is that GIS raises new issues that traditional archaeologists could not deal with. We consider that if GIS-based methods are fused with methods in archaeology and those of the humanities and social sciences, such approaches can deal with two fundamental information types, space and time data for entities in every material culture, and they will be developed into spatio-temporal science.

For instance, consider the case in which traditional archaeologists discovered a pottery style. They tended to discuss the historical and regional characteristics of a material culture based on the relative position of the pottery in relation to those of similar potteries and remains. However, because the pottery did not have absolute spatio-temporal coordinates, it was almost impossible to compare it in a rigorous context with that from completely different cultures and regions. What the traditional archaeologists made was an analogical inference based on subjective interpretation without considering natural, ecological, and environmental situations of the same age. GIS-based spatial archaeology can overcome these limitations.

As shown in this chapter, the Sannai-Maruyama archaeological site project has produced various interesting new results about "human–nature interactions." We obtained the following results concerning analytical techniques.

First, drill cores and the derived geological-column data are useful in the reconstruction of the DEM of previous times using a spatial-interpolation method.

Second, the introduction of a linear time scale, such as that derived through calibrated radiocarbon dating in three-dimensional geological structures, is indispensable in order to understand the geomorphological characteristics of a specific period.

Third, the Boolean-overlay method is valuable in estimating relative values of productivity when many uncertain factors exist.

Fourth, the slope-adding method, with respect to radial direction form of a site, is useful for the approximate estimation of areas contained within a certain time radius of walking.

We conclude in the historical interpretation that the two distinctive characteristics of the Sannai-Maruyama site are a very large and long-lasting settlement, which is likely to be explained by a large site-catchment area that provided abundant vegetation. The intensive horticultural subsistence around Sannai-Maruyama can be considered as a prerequisite to the fostering of subsequent agriculture.

In conclusion, we consider that GIS-based spatial archaeology is able to integrate various types of spatio-temporal information in archaeological analysis, and this ability provides new approaches and insights.

Acknowledgments

I would like to acknowledge that this work was done in the framework of the Sannai-Maruyama archaeological site project (II), (III), founded by the board of education of Aomori Prefecture. I would like to thank Professor Sei-ichiro, Tsuji (The University of Tokyo), and Professor Toyohiro, Nishimoto (National Museum of Japanese History), coordinator of these projects, for their support. And I wish to thank my adviser Sumiko, Kubo (Waseda University), Takashi, Oguchi (The University of Tokyo), Masakazu, Tani (Kyushu University), and Takeji, Toizumi (Waseda University), for their assistance with my research.

References

Allen, K.M.S., Green, S., and Zubrow, E., Eds., *Interpreting Space: GIS and Archaeology*, Taylor & Francis, London, 1990.

Ceccarelli, L. and Niccolucci, F., Modelling time through GIS technology: the ancient Prile Lake (Tuscany, Italy), in *The Digital Heritage of Archaeology*, Hellenic Ministry of Culture, Greece, 2003, pp. 133–138.

Clarke, D.L., *Spatial Archaeology*, Academic Press, London, 1977.

Clevis, Q., Tucker, G., Lock, G.R., and Desitter, A., Modelling the Stratigraphy and Geoarchaeology of English Valley Systems, English Heritage — ALSF Report, 2004.

Crescioli, M., D'Andrea, A., and Niccolucci, F., A GIS-based analysis of the Etruscan cemetery of Pontecagnano using fuzzy logic, in *Beyond the Map: Archaeology and Spatial Technologies*, IOS Press, 2000, pp. 157–179.

Doortje, V.H., Agency and GIS: the Neolithic land use hypothesis within Southern Italy, in *The Digital Heritage of Archaeology*, Hellenic Ministry of Culture, Greece, 2003, pp. 201–207.

Gaffney, V. and Stančič, Z., GIS approaches to regional analysis: a case study of the island of Hvar, Ljubljana: Znanstveni institut Filozofske Fakultete, 1991.

Hatzinikolaou, E., Hatzichristos, T., Siolas, A., and Mantzourani, E., Predicting archaeological site locations using GIS and Fuzzy Logic, in *The Digital Heritage of Archaeology*, Hellenic Ministry of Culture, Greece, 2003, pp.169–177.

Hodder, I. and Orton, C.R., *Spatial Analysis in Archaeology*, Cambridge University Press, Cambridge, 1976.

Indruszewski, G., Reconstructing the seascape at the mouth of the Oder: Elaboration of a DBM–model based on 1912-soundings, in *Archaeological Informatics: Pushing the Envelope*, Archaeopress, Oxford, 2002, pp. 63–70.

Kamermans, H., Land evaluations as predictive modelling: a deductive approach, in *Beyond the Map: Archaeology and Spatial Technologies*, IOS Press Amsterdam, Washington, DC, Tokyo, , 2000, pp. 124–146.

Kaneda, A., Tsumura, H., and Niiro, I., *Geographical Information System for Archaeology* Kokon-shoin, Tokyo, 2001, (in Japanese only).

Lock, G.R. and Stančič, Z., *Archaeology and Geographical Information Systems: A European Perspective*, Taylor & Francis, London, 1995.

Renfrew, C. and Bahn, P., *Archaeology: Theories, Methods and Practice*, 3rd ed., Thames & Hudson, New York, 2000.

Reynoso, C. and Castro, D., Chaos and Complexity Tools for Archaeology: State of the Art and Perspectives, in *Abstracts Book of CAA2004 Conference — Beyond the Artifact*, CAA Italy, Prato, 2004, p. 84.

Reynoso, C. and Jezierski, E., A Genetic Algorithm Problem Solver for Archaeology, in *Archaeological Informatics: Pushing the Envelope*, Oxford: Archaeopress, Oxford, 2002, pp. 507–510.

Spikins, P.A., GIS models of past vegetation: an example from Northern England, 10000–5000 BP, *J. Arch. Sci.*, 27–3, 219–234, 2000.

Tsumura, H., A new analysis technique of reconstructing and assessment of paleo environment for archaeological studies, Proceeding of Asia GIS, Center for Spatial Information Science, the University of Tokyo, 2001.

Tsumura, H., An assessment of the inter-site conjunctions by A-index: through a case study on the Neolithic settlement (Aomori, Japan), in *Enter the Past — The E-Way into the Four Dimensions of Cultural Heritage*, Archaeopress, Oxford, 2004.

Verhagen, P. and Berger, J.F., The Hidden Reserve: Predictive Modelling of Buried Archaeological Sites in the Tricastin–Valdaine Region (Middle Rhone Valley, France), in *Computing Archaeology for Understanding the Past*, Archaeopress, Oxford, 2001, pp. 507–510.

14

Migration, Regional Diversity, and Residential Development on the Edge of Greater Cairo — Linking Three Kinds of Data — Census, Household-Survey, and Geographical — with GIS

Hiroshi Kato, Erina Iwasaki, Ali El-Shazly, and Yutaka Goto

CONTENTS

14.1 Introduction

In Egypt, major social problems occur in urban areas, because the areas absorb people who move from rural areas to find work. To understand the urban–social problems, it is also necessary to understand rural societies, because the social problems in urban areas reflect the rural transformation. Therefore, migration that relates both urban and rural sectors is an apposite topic in the study of Egyptian society. However, it was fairly difficult to carry out the studies of Egyptian migration that considered the linkage between urban and rural societies. Major reasons are the difficulty in conducting a survey and the lack of reliable maps. Also, military and security restrictions hampered empirical studies based on microdata and maps. Fortunately, these constraints lessened after the introduction of the open-economic policy in the 1980s. Surveys in urban and rural areas are now becoming easier (Datt, 1998; Nagi, 2001; Assad, 2002; Government of Egypt, 2002), and there is a better environment in which the impact of migration can be examined by using many different materials and methods. Taking advantage of this better environment, this chapter shows how to integrate various macrodata and microdata with Geographical Information Systems (GIS) to analyze migration behavior in Egypt.

This chapter is composed of five sections, including this introductory section. The next section, Section 14.2, introduces the sources of data and methodology for the survey. Section 14.3 analyzes determinants of migration from rural areas to two survey areas in urban areas. Section 14.4 examines residential developments resulting from migration. The chapter ends in Section 14.5, summarizing major results with remarks on future studies.

14.2 Data and Methodology

14.2.1 Data

This chapter links three kinds of data: macrodata, mainly provided by the population census; household-survey microdata; and geographical data.

Statistical data on modern Egypt are relatively abundant, as the population census has been published almost every 10 years from the end of the 19th century (CEDEJ, 2004). The linkage of these macro statistical data with geographical data from maps, as well as the collection of household-survey facts, gives originality to this study. Central Agency for Public Mobilization and Statistics (CAPMAS) has a department of GIS working to digitize maps that display various facts about Egypt. For example, the administrative digital map at the village level available from this center enables the integration of the census and microdata with GIS data at village level.

It is now possible to view concurrently both the rural areas and Greater Cairo, which are connected through migration, and analyze the spatial patterns of migration and settlement.

Geographical data, such as location of the apartments where each household resides, as well as migrants' home villages, was collected during fieldwork. We personally undertook the collection of geographical data for this study because of the following two reasons.

The first is technical and is related to the sampling problem. As there is no information available on the birthplace of residents, it was necessary to check all families living in the survey areas to find those household heads who had been born in rural areas. Information was also needed on where the target homes were situated, and the street names, blocks, buildings, and floors. As will be discussed, the surveyed areas are relatively new developments without urban planning and, without the fieldwork information, it would have been impossible to identify the target households.

The second reason is related to the research interest. Data on the sources of rural migrants and where they have settled are essential indicators of migration patterns.

The collected geographical data was coded and attached to the digital map provided by CAPMAS. A building-level digital map of the surveyed areas was also prepared using this base map as a reference, as described below.

14.2.2 Selection of the Survey Areas

As is well known, Egypt is a typical hydraulic society based on the Nile River, which divides into two main branches and forms a delta just north of Cairo. Egypt is administratively composed of two regions: Lower Egypt is the northern part of the country from Cairo to the Mediterranean, and Upper Egypt is the southern part from Cairo to the border between Egypt and Sudan. Each region is divided into three hierarchical divisions: governorate, *qism*, and *shiyakhat* for urban sectors, and governorate, *markaz*, and *qarya* (village) for rural regions. The four large cities of Cairo, Alexandria, Port Said, and Suez are counted as individual governorates. The smallest unit of these urban governorates (as well as urban centers of other governorates) is the *shiyakhat*; *qarya* is the smallest unit for rural parts.

Taking the concern of this chapter into consideration, two areas on the edge of Greater Cairo were selected for case studies, based on the criteria of being low-income, residential areas housing rural migrants. On the northern edge is *shiyakhat* Bigam in *qism* Shobra El-Kheima, and on the southwestern edge lies *shiyakhat* Zinin in *qism* Bulaq El-Dakrur (Figure 14.1).

Although the smallest administrative unit, a *shiyakhat* covers quite a large area: Bigam is 7,154,459 m^2 and has a population of 336,957 (1996), while Zinin covers an area of 1,140,279 m^2 and has a population of 106,957 (1996).

Since it is impossible to cover the whole of a *shiyakhat*, the survey areas were chosen at block level, on the edge of the administrative border. The survey blocks in Bigam lie on the border with *qarya* Manta (*markaz* Kalyoub), and are located near the industrial zone of Shobra El-Kheima (Figure 14.2). The survey blocks in Zinin lie on the border with *shiyakhat* Bulaq El-Dakrur,

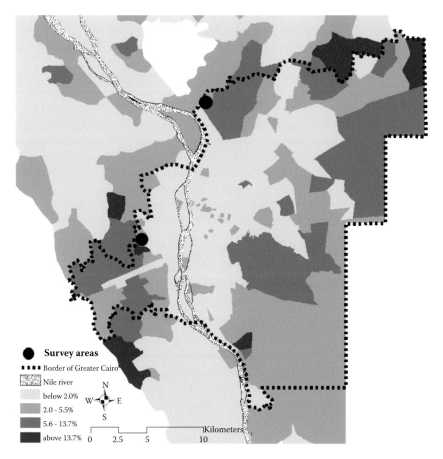

FIGURE 14.1
Location of the survey areas in the map and percentage of buildings under construction by *shiyakhat* (Greater Cairo).

close to the commercial/residential districts of El-Dokki and El-Giza. As shown in the figure, the areas are adjacent to vacant land, which indicates the recent transformation of agricultural fields to developed urban residential areas. Indeed, the two areas are typical of those on the edge of Greater Cairo, which developed rapidly from the late 1970s.

14.2.3 Data Sampling

The original microdata used in this chapter were collected during the household survey undertaken by the Graduate School of Economics, Hitotsubashi University, in collaboration with CAPMAS during the years 2002 and 2003 (Kato, 2004).

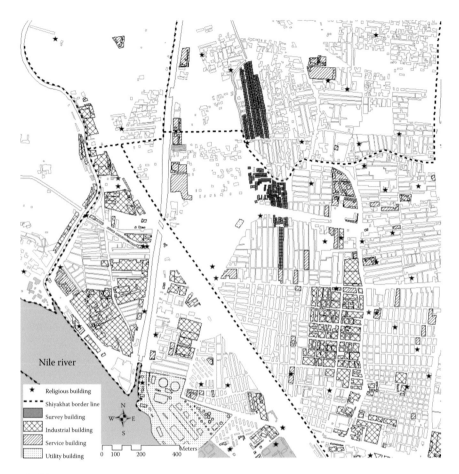

FIGURE 14.2
Location of the two survey areas (Bigam).

The problem inherent in a survey of migrants is the sampling. Since there is no information available on how many migrants are in Greater Cairo or where they live, it is impossible to establish a sample group that represents the entirety of rural migrants living in the city. For this reason, the survey was conducted as a case study in selected geographical areas.

The survey collected information from two household categories: those whose household heads had moved directly from the rural areas to Greater Cairo, and those with heads of households born in Greater Cairo. Since most of the household heads living in the survey areas are believed to have been born in Greater Cairo, the samples of those born in that city can be considered as representing the majority of household heads in the survey areas. The data of those born in Cairo were used in part to clarify the characteristics of the migrants in the survey blocks.

TABLE 14.1

Samples

		Bigam		Zinin		Total	
		%	number	%	number	%	number
Households in the	Rural Migrants	32.5	635	23.4	627	27.2	1262
Sampling Survey	Other	67.5	1321	76.6	2055	72.8	3376
	Total	100.0	1956	100.0	2682	100.0	4638
Samples Surveyed	Rural Migrants		400				800
	Born in Greater Cairo		200				400
	Total		600				1200

Before the household survey, a preparatory sampling survey was held to collect information from the apartment-building owners on (1) household heads who came directly from rural areas to Greater Cairo, and (2) the location of each household's residence (street, building, and floor). The survey started from a block on the administrative border of the two *shiyakhat*, selected randomly, and the survey moved on to the neighboring blocks one after the other from any direction until the intended number of samples in each of the two survey areas was met. Each block was surveyed starting from one of its corners, and the block was kept on the right-hand side of the surveyors as they moved from one apartment to the next.

14.3 Migration and Regional Categorization

By far the largest number of migrants to Cairo emanates from the governorates of Menoufia and Suhag, in Lower and Upper Egypt, respectively, but the nature of their migration patterns is quite different. The following detailed analysis of the regional characteristics at the village level reveals the reasons for the difference.

The geographical data determined that migration characteristics were of three types:

1. Where the migrants come from.
2. The determinants of out-migration to Greater Cairo.
3. Regional categorization with a focus on income and employment structures within a region.

Three sources of information available were the information on locations of the migrants' home villages from the household-survey data, the data from the population census, and geographical data from the digital map of Egypt.

14.3.1 Migrants' Villages of Origin

The maps of the migrants' home villages indicate the linkage of certain rural areas with the survey areas (Figures 14.3 and 14.4). In Upper Egypt, the governorate of Suhag sent out relatively more migrants to Zinin, whereas in Lower Egypt, the governorate of Menoufia provided more migrants for Bigam.

The maps also indicate the concentration of migrants' originating areas not only at the governorate level but also at the *markaz* level. The migrants' home villages in Menoufia are concentrated in the south, while those in Suhag are concentrated in the north.

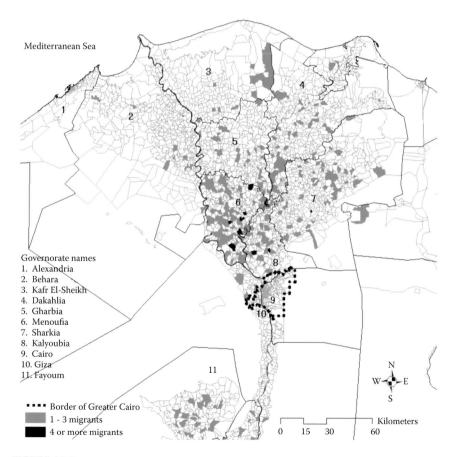

FIGURE 14.3
Number of migrants by village of origin (Lower Egypt).

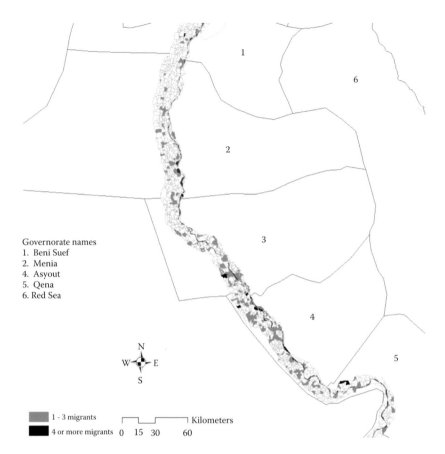

FIGURE 14.4
Number of migrants by village of origin (Upper Egypt).

14.3.2 Characteristics of the Regions of Origin

To understand the characteristics of the migrant source areas, three levels of analysis were conducted, based on the hypothesis that the determinants of out-migration differ by region. This hypothesis is drawn from the mapping by village of the major indicators, such as population density, income level, age, education level, unemployment, and the employment situation. The mapping procedure enabled the development of a hypothesis that the income level and job opportunity at home are the important factors.

The first analysis conducted was the logit estimation of determinants to show the factors that determine out-migration (Table 14.2).

The probability of out-migration is described by a dummy variable either having (= 1) or not having a migrant(s) (= 2) in a *markaz* or *qarya*. The estimation was done on *markaz* level for the analysis for the whole of Egypt,

TABLE 14.2

Determinants of Out-Migration to Greater Cairo (logit estimation)

	Markaz Level Whole Egypt		Village Level Lower Egypt		Village Level Upper Egypt	
	coefficient	z-statistics	coefficient	z-statistics	coefficient	z-statistics
Distance to Cairo	-0.002	-1.48	-0.025	-7.36***	0.002	2.98***
Population density	0.326	2.73***	—	—	—	—
GDP per capita	-0.001	-4.05***	—	—	—	—
Unemployment rate	0.049	2.39**	0.012	0.66	0.028	1.42
Proportion of workers in	0.143	2.15**	0.018	2.12**	-0.005	-0.66
Proportion of workers in private sector	-0.056	-1.69*	-0.035	-3.00***	0.009	0.81
Unemployment rate in Markaz town	—	—	-0.046	-1.46	0.084	2.53***
Constant	5.710	2.01**	1.661	2.26***	-4.137	-4.79***
Pseudo R-squared		0.34		0.105		0.039
(pr>chi-squared		(0.000)		(0.000)		(0.000)
n		151		1830		1261

Notes: *** indicates statistical significance at 1% level, ** at the 5% level, * at the 10% level.

Urban *markaz* (*markaz* composed exclusively of *shiyakhat* and *qism*) are excluded from the analysis, since we treat here the rural migrations.

The distance is measured between the markaz town (or in case of its absence, central point in *markaz*) to Cairo center (Taharir square).

Source: From population census 1996, household survey data.

and on *qarya* level for the analysis by region, due to the nature of samples and data.

Per capita, Gross Domestic Production (GDP) (LE/year, 2000/2001) was used as the indicator of income level. Two sets of variables were employed with regard to employment opportunities. One was the proportion of workers in the agricultural sector taken together with the population density to estimate the employment opportunity in the agricultural sector. The other was the unemployment rate compared with the proportion of workers in the private sector to estimate the employment opportunity in the nonagricultural sector (agricultural employment in the private sector is controlled by the proportion of workers in the agricultural sector). It is assumed that the greater the proportion of workers in the private sector, the larger will the nonagricultural employment opportunity become.

The second analysis was the estimation of income levels (Figures 14.5 and 14.6), and the third was the cluster analysis of the employment structure in rural Egypt.

Major indicators of employment contained in the population census (1996) were used to clarify the regional diversity of employment structures

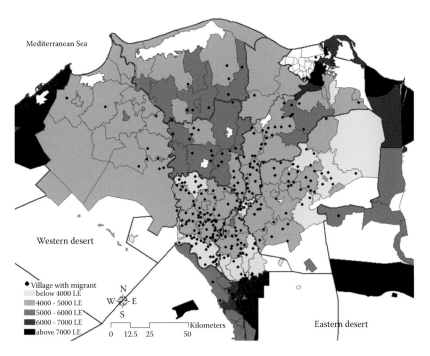

FIGURE 14.5
Income level of *markaz* sending out migrants (Lower Egypt).

in Egypt. In terms of this, the villages of Egypt are categorized into following four groups.

Group 1: Characterized by the predominance of the government sector and high unemployment

Group 2: Distinguished by private sector, nonagricultural activities

Group 3: Noted by the size of the agricultural sector

Group 4: Determined by the size of the agricultural sector, but predominantly composed of the self-employed (Table 14.3)

The villages in Lower Egypt are classified into two types: the areas within Menoufia that belong to Group 1, and the outer zone areas belonging to Group 4. The migrants' sourcing villages are concentrated in the zone composed of Group 1 (Figure 14.7). On the other hand, villages in Upper Egypt are divided into two zones: one to the north of Menia belonging to Group 4, and another to the south of Menia, notably of governorates, such as Suhag, and the southern parts of Asyout and Qena. The latter zone has villages with a more diversified employment structure belonging to Groups 1 and 2. It is

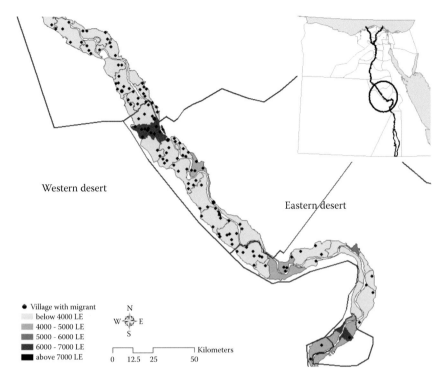

FIGURE 14.6

Income level of *markaz* sending out migrants (Upper Egypt).

TABLE 14.3

Summary Statistics of Cluster Groups (Wards Method)

	Group 1	Group 2	Group 3	Group 4
Number of clusters	2074	517	1477	404
Unemployment rate in markaz town (%)	10.9	9.2	11.1	10.4
Unemployment rate in village (%)	11.7	6.0	7.7	5.1
Proportion of workers in private sector (%)	65.6	76.4	83.7	88.8
Proportion of workers in public sector (%)	3.7	6.0	1.2	0.7
Proportion of workers in government sector (%)	29.9	16.4	14.1	9.2
Proportion of workers in agriculture (%)	41.1	33.8	69.0	72.1
Proportion of workers in manufacturing (%)	10.6	15.6	4.9	3.3
Proportion of workers in commerce (%)	4.9	8.0	3.1	3.5
Proportion of workers in public administration & defense (%)	10.7	5.6	5.6	3.5
Proportion of workers in education (%)	11.4	5.6	5.4	3.9
Proportion of workers in health & social works (%)	2.2	1.2	0.9	0.6
Proportion of workers in construction (%)	5.4	16.3	2.8	3.2
Proportion of self-employed (%)	22.1	21.4	20.8	50.8
Proportion of waged (%)	58.5	64.9	51.9	33.3

Source: population census 1996.

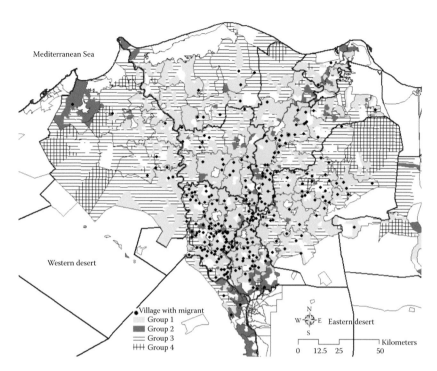

FIGURE 14.7
Migrants and village categories (Lower Egypt).

these governorates that send out most of the migrants in Upper Egypt (Figure 14.8).

These analyses are done on a *qarya* and *markaz* level, and are linked with the geographical information through mapping. The digital map also served to determine the distance factor in the out-migration analysis.

The results can be summarized as follows: The migrants' source areas have the common traits of a low-income level and a low employment opportunity, but with some regional differences. In Lower Egypt, migrants come from villages located in the proximity of Greater Cairo that offer fewer job opportunities in the nonagricultural private sector. In Upper Egypt, the migrants come from villages located far from Greater Cairo, regardless of the employment structure. They tend to leave their villages when the job opportunity in the nearby towns is small, whereas the migrants of Lower Egypt leave their villages regardless of the job opportunity in the nearby towns. This may be due to the distance factor, and the extremely low level of income, and may possibly be related to the employment opportunity being limited to construction and other service sectors.

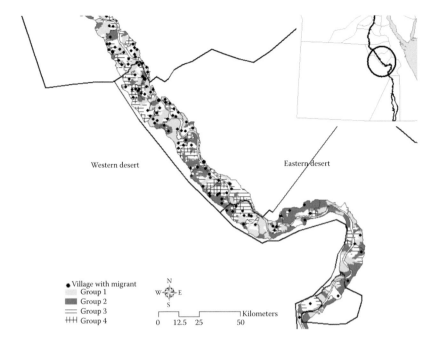

Western desert

Eastern desert

• Village with migrant
   Group 1
■ Group 2
= Group 3
╫ Group 4

N
W-⊕-E
S

Kilometers
0   12.5   25   50

FIGURE 14.8
Migrants and village categories (Upper Egypt).

14.4 Migration and Residential Development

14.4.1 Constructing the Building Map

Bigam and Zinin, where the two survey areas are located, are *shiyakhats* with a low income level that also absorb people from outside and are developing, as shown in Figure 14.1 — buildings under construction. The two survey areas are typical developing areas on the edge of Greater Cairo.

The next concern in this chapter is to examine the ways the two survey areas have developed and the relationship between the settlement patterns of rural migrants and the residential development of the two areas. The first step in this concern was to make detailed residential maps of the survey areas. The CAPMAS GIS map covering blocks does not contain sufficient information to study those developing edges of Greater Cairo, such as the survey areas.

The map was constructed in the following five steps:

1. Measurement of the dimensions of each building

2. Digitizing and coding each building to form new polygon shape files, and overlaying these on the block map of CAPMAS
3. Attributing the building data to each polygon representing a building
4. Digitizing and coding each household to form a new dot shape file, and overlaying these on building polygons
5. Attributing the household data to each dot representing the household

14.4.2 General Description of the Survey Areas

The detailed geographic information collected from the survey on 1) block division, 2) building heights, and 3) building occupiers can clarify the informal development of the survey blocks. The three-dimensional shape of blocks varies in size. In Bigam, one zone of 19 blocks is divided into 290 buildings, compared to another zone of only six blocks but with 405 buildings. Zinin shows a similar contrast, with 496 buildings in the nine blocks of one zone, yet with 18 blocks containing only 384 buildings in another zone (Figure 14.9, and photograph in Figure 14.10). This is because of the manner of block development, which starts as a free-standing building in open land and then develops haphazardly into attached houses or scattered buildings.

Building heights vary, destroying the uniform relationship between the very narrow streets of only 3 meters in width and buildings that vary from eight floors (25 m tall) to low buildings of one to two floors, and the façades remain unfinished. Furthermore, the infrastructure has relied on the personal efforts of occupiers digging wells into the water table and making sewage trenches. There are no playgrounds or any public space for recreation. The buildings of this informal settlement continue to develop without the compulsory licensing required by the government.

14.4.3 Settlement Patterns and Residential Development

As GIS methodology allowed us to achieve a linkage between the microdata collected from the household survey and the digital maps, the settlement patterns of rural migrants and the residential development of the survey areas can be observed in relation to each other.

The examination of residential development is crucial to the study of settlement patterns, since the choice of a residence may also be determined by its availability. The settlement patterns will thus be considered in terms of the occupiers' region of origin and other attributes using the information on the residential location at the building/apartment level.

According to the interviews, it was from the late 1970s or early 1980s that large numbers of people flowed into the survey blocks. Interviewees (NGO

FIGURE 14.9 (a and b)
Survey blocks (Bigam).

FIGURE 14.10 (a and b)
(See color insert following page 176.) Typical buildings in the survey block (left: Bigam/right: zinim).

directors, long-standing residents) unanimously point to two factors for this: the international migration to oil producing countries, and the open door policy (*Infitah*).

In reality, most of the migrants and those born in Greater Cairo moved into their residences in the survey blocks after the 1980s. This increase would be related to the residential development of the survey blocks. The proportion of occupiers renting the residences increased around the 1970s or 1980s, which implies that many apartments were rented out.

The rental apartments appear to cluster in certain buildings (Figure 14.11a) and seem to be those where the current residents moved to recently. It should be noted that this phenomenon is not due only to the residential development of the survey blocks, nor is it because the apartments became available in the most recent decades. Even among the migrants who acquired or rented their residence after the year 1980, earlier migrants are more likely to own the residence, whereas recent migrants are more likely to rent the residence (Figure 14.12).

The difference in behavior in the acquisition of the residence may be related to the difference of migrants' trajectories after arriving in Greater Cairo from their home villages. There is a considerable time lag between the year of arrival in Greater Cairo from the home village and the estimated year of settlement into the residence. The time lag is greater for those migrants who arrived in Greater Cairo in the early period, some of whom moved into residence in Bigam and Zinin after accumulating money to buy their apartment.

Figure 14.11b indicates the settlement patterns of migrants by their region of origin. It seems that migrants prefer to settle into the building or a neighboring building, where migrants from the same region of origin live. In particular, migrants from Menoufia seem to have this tendency.

14.5 Conclusions

Previous studies of migration, as well as studies of income and the structures have treated Egyptian rural society as a homogeneous entity, or at most have divided the country into the two regions of Lower and Upper Egypt, both in the past and at present. Contrary to this image, an analysis that combines statistical data on migration and regional categorization with GIS maps suggests a different outlook.

As the regional categorization by employment structure at village level suggests, the job-opportunity situation in rural areas seems to be diverse in size and nature. This diversity, as well as the level of income, is one of the determinant factors that affect the rural–urban migration linkage in Egypt. The phenomenon of rural migration to Greater Cairo is linked not only to the urban labor market but also to the local, rural-area labor market.

FIGURE 14.11 (a and b)
Ownership of residence (a) and region of origin (b) (Bigam).

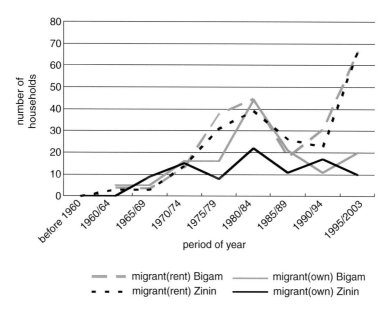

FIGURE 14.12
Settlement period of migrants to the actual residence.

This diversity of job opportunity may explain, in part, the impossibility of the Cairo labor market to offer sufficient jobs under the transitional economy, on one hand (Radwan, 2000; El-Laithy, 2003), and the possible formation and development of new labor markets in local provinces, on the other. Thus, it is concluded that in order to understand the rural–urban linkage in Egypt, more detailed study is needed to clarify the diverse characteristics of the local, rural-area labor market, as well as that in Cairo.

It is apparent that GIS is an efficient tool for the clarification of the migration from rural areas and the informal development of suburban areas in Greater Cairo. In parallel with the building map, the GIS method can be further adapted to study the urban characteristics of Bigam and Zinin and to visualize the developmental processes in the survey areas.

The following two issues, especially, are expected to be important in the study of the survey areas. The first is the urban structure that considers how the network of streets and open spaces is connected, where the streets are categorized according to their widths. The second is the building style, noting the method of construction, time of building development, number of floors, façade materials, and the provision of infrastructure.

References

Abu-Lughod, J., "Migrant Adjustment to City Life: the Egyptian Case", *American Journal of Sociology*, vol.67, n.1, 1961.

Assad, R., Ed., *The Egyptian Labor Market in an Era of Reform*, Cairo: The American University in Cairo Press, 2002.

CEDEJ, Cédérom Interactif, "Un siècle de recensements en Egypte (1882-1996)", http://www.cedej.org.eg, 2004.

Datt, G. et al., "A Profile of Poverty in Egypt: 1997", *FCND Discussion Paper*, n.49, Washington D.C.: IFPRI, 1998.

Deboulet, A., "Etat, squatters et maîtrise de l'espace au Caire", *Egypte: Monde Arabe*, no.1, 1990.

El Kadi, G. *L'urbanisation spontanée au Caire*, Fascicule de Recherches, no.18, Tours, 1987.

El-Laithy, H., "Poverty and Economic Growth in Egypt, 1995-2000", *Policy Research Working Paper*, 3068, Washington D.C.: The World Bank, 2003.

Esfahani, H.S., "Aggregate Trends in Four Main Agricultural Regions in Egypt, 1964-1979", *International Journal of Middle East Studies*, vol.20, no.2, 1988.

Government of Egypt & The World Bank, *Arab Republic of Egypt: Poverty Reduction in Egypt, Diagnosis and Strategy*, 2 vol., Washington D.C.: World Bank, Report n.24234-EGT, 2002.

Ireton, F., "The Evolution of Agrarian Structures in Egypt: Regional Patterns of Change in Farm Size", Hopkins, N. & Westergaard, K, Eds., *Directions of Change in Rural Egypt*, The American University in Cairo Press, 1998.

Kato, H., Iwasaki, E., and El-Shazly, A. "Internal Migration Patterns to Greater CairoÅ|Linking Three Kinds of Data: Census, Household Survey, and GIS ", *Mediterranean World*, 17, the Mediterranean Studies Group, Hitotsubashi University, Tokyo, 2004.

Nagi, S. Z., *Poverty in Egypt: Human Needs and Institutional Capacities*, Lexington: Lexington Books, 2001.

Radwan, S., "Employment and Unemployment in Egypt", ECES, *Working Paper*, n.70, Cairo, 2000.

15

Effect of Environmental Factors on Housing Prices: Application of GIS to Urban-Policy Analysis

Yasushi Asami and Xiaolu Gao

CONTENTS

15.1 Economic Value of Residential Environment

Objectively showing the propriety of a policy is more and more recognized to be important in the process of policy-making. However, in many cases, this is difficult because of the lack of effective evaluation methods for policies. Taking city-planning systems as an example, so far the procedures of planning have only been justified in view of the physical needs of environmental improvement and in view of the legitimacy of planning procedures. Yet there is no procedure for confirming the appropriateness of a policy based on an objective, quantitative evaluation of the policy, although this is agreed to be very important.

Many efforts have been made to evaluate the effect of urban policies. For example, many studies analyzed the effect of zoning ordinance by applying a hedonic approach (Mark and Goldberg, 1986; Pasha, 1992; Pogodzzinski and Sass, 1991; Schilling, Sirmans, and Guidry, 1991). However, because of the use of general data, the accuracies of analyses are often low, and the results only showed the overall trends or effects of policies. Therefore, the implications of the analyses are not reliable enough to be applied to policy-making procedures. Recently, this situation has largely been changed. Various detailed, high-resolution spatial information becomes available with the rapid development of geographical information systems (GIS), and, accordingly, precise analyses employing these data become possible. In this chapter, an example of evaluating city-planning policies is given. In particular, the usefulness of detailed spatial data in the analysis of microlevel residential environment is illustrated, based on which policies concerning the improvement of residential environment are addressed.

Turning to urban policy issues in Tokyo, the size of sites for detached houses in Tokyo has been decreasing year by year. Tokyo Metropolitan Government (2003) reports that average the area of land owned by individuals in Tokyo's 23-ward region decreased from 284 square meters in 1978 to 211 square meters in 2002. It also reports that land less than 200 square meters is more than 77 percent of all land owned by individuals in Tokyo's 23-ward region. Continuing demand for land in this region surely accelerates the decreasing tendency of lot size. In particular, since the remarkable rise of land price in the 1980s, subdivision phenomenon in residential areas has been prevalent. Density increase led to various problems that aggravated the residential environment, for example, it caused the decrease of greenery, obstruction of sunshine, insufficient ventilation environment, and, in the event of fire, it increased the danger of fire expansion.

A housing-demand survey in 1993 reported that 45 percent of households living in detached houses were dissatisfied with access to public green space, 36 percent were dissatisfied with residential environment, and 32 percent were dissatisfied with sunshine and ventilation. The improvement of residential environments has become an urgent issue to be resolved in urban policy.

Land value of detached houses has been extensively analyzed from various points of view. However, analyses developed by existing studies mainly paid attention to the value-deciding factors of the target sites, and not much analysis had been made for the external effects with regard to surrounding sites.

Although a number of regulating and inducing methods were proposed for improving the microlevel residential environment (Special Committee for Research on Urban Housing in the Architectural Institute of Japan, 1996; Kuwata, 1998), under the conventional framework of city planning, neither evaluating the proposals nor choosing appropriate countermeasures based on the analyses of social effects is possible.

This chapter tackles such evaluation issues with a microeconomic approach. Several literatures have made similar trials. Li and Brown (1980), for example, estimated the effects of townscape, noises, and distance to factories and shopping areas and showed that estimating housing prices without incorporating these microlevel environmental factors is subject to statistical biases. Kameda and Hidano (1997) analyzed the external effect of greenery on residential environments. Their empirical results allowed for cost-benefit analyses for improvement projects. Hidano, Kameda, and Ando (1998) quantitatively analyzed the joint effect of planting trees and building setbacks, which provides space for the trees, and calculated the benefits and costs for each residential site, which was applied to make a fair cost-sharing plan in the neighborhood-improvement project. These studies, however, are still limited in that many effects, such as the broadening of adjacent roads, increase in sunshine and ventilation, and change in the size of gardens, were ignored, though they should also affect the residential environment. Variables on sunshine duration, accessibility to greenery, situation of gardens, and so on, will be incorporated into the following analysis to find out their economic effects, as well as their external effects, on housing prices. In turn, the analyses based on the results imply effective measures for improving residential environments so as to realize efficient land use in detached-housing areas.

15.2 Method and Data

Hedonic approach is effective to calculate external and social effects. Based on the capitalization hypothesis that the improvement in environment is capitalized into land price, the benefits of environmental factors can be estimated by analyzing land price. In a perfectly functioning market, households that will maximize their utility purchase houses such that their willingness to pay for a factor of residential environment equals its shadow price, or hedonic price. In equilibrium, the hedonic price for each factor becomes equal to its marginal value. Therefore, by estimating their hedonic prices, the value for the factors of residential environments can be estimated. Such information can be used, for example, to evaluate changes in consumer surplus due to a project that changes residential environments (Boardman, et al., 1996).

Since the hedonic approach was economically formulated by Rosen in 1960s, it has been frequently used in project analyses and policy evaluation, such as the evaluation of deterioration in residential environments (Kain and Quigley, 1970), house price index in metropolitan areas (Goodman, 1978), amenity for urban houses (Blomquist and Worley, 1981), residential environments (Kanemoto, Nakamura and Yazawa, 1989; Yazawa and Kanemoto, 2000), regional housing-price analysis (Mills and Simenauer, 1996), and the effects of regional and cultural characteristics on housing price (Huh and

Kwak, 1997). Li and Brown (1980), Kameda and Hidano (1997), and Hidano, Kameda, and Ando (1998), referred to in the former section, also used the hedonic approach.

The sales data of detached houses with land sites, which were listed in the *Weekly Housing Information* magazine from October 1996 to September 1997, are used for the current analysis. The prices when the entry for a property is erased from the list were used, with the assumption that they are close to the actual prices. Using such data instead of actual prices may cause an upward bias, because the bargain on price may succeed in the final course of transaction and, in addition, most of the listed properties are free from illegal construction. Nonetheless, the qualitative result will remain the same even with these biases under the current market situation.

Location factor represented by the distance to central business districts (CBD) may mask the effects of local factors, because this factor is very strong in influencing the property prices. Hence, location factor should be removed as much as possible to analyze local factors (Goodman, 1978; Cropper, et al., 1993). For this effect, the study area was confined to the area along one railway line. After careful examination, 190 properties were selected for the analysis, which lie in the territories of five stations along Odakyu line in Setagaya Ward, Tokyo. Odakyu line extends westward from Shinjuku station, one of the largest CBDs in Tokyo. The study area is between 17 km and 23 km from the center of Tokyo, and all the properties are in the first category, exclusive, low-rise, residential zone (purely used for low-rise houses).

Data that possibly influence the value of house and lot were collected for these properties. The data can be classified into two groups: regional factors describing the regional characteristics and individual factors describing the specific characteristics of the houses and lots. Efforts were made to collect individual factors as much as possible in this research.

Through an on-site survey, the individual data unavailable in the *Weekly Housing Information* database were collected, including the actual land use, the distance between neighboring buildings, and a variety of local residential environments. In addition, the sample properties were matched on 12,500 residential maps, and their distances to the nearest schools and hospitals were measured with GIS.

The sunshine duration of each property was calculated using sunshine-duration software (EX-Shadow by A&A Co., Ltd.) based on the three-dimensional Computer Aided Design (CAD) models of buildings upon the residential maps of Zenrin Co., Ltd. and the building information obtained from on-site survey. To simplify the process, a 3-meter height for each floor was assumed, and the sunshine duration was calculated on the 1.5-meter plane above the average ground level, at the most advantageous sunlight-acquisition point on the external walls. Besides, ventilation status was estimated for each house in terms of its distance to adjacent buildings. Table 15.1 summarizes the collected data. For details of the data, see Gao and Asami (2001a).

TABLE 15.1

Descriptive Statistics of Variables

Variables	Definition (Unit)	Min	Max	Average	Standard Deviation
Actual_FAR	Building floor area / lot area (ratio)	.35	1.83	.9608	.2650
Build_quality	Dummy of good quality of buildings in the local district, if true, 1	.00	1.00	.3632	.4822
Greenery	Dummy of adjacent to public green space, if true, 1	.00	1.00	6.316E-02	.2439
Greenery/S		.00	.02	5.229E-04	2.318E-03
Frontage	Frontage of lot (m)	1.00	20.00	7.4274	3.6023
Good_ pavement	Dummy of good pavement of front road, if true, 1	.00	1.00	.5105	.5012
Landscape	Dummy of within designated landscape areas, if true, 1	.00	1.00	.2421	.4295
Lot_area	Lot area (m²)	34.72	588.05	124.6675	81.2435
Mixed_use	Dummy of non-residential buildings more than 1/3 in the local district, if true, 1	.00	1.00	.1105	.3144
Mixed_use/S		.00	.02	1.373E-03	4.043E-03
P	Market-clearing price of land with house (million yen)	31.50	398.00	94.5846	60.5114
P/S		.46	1.24	.7816	.1853
Parking_lot	Number of parkable cars (cars)	.00	3.00	1.2526	.7413
Residual_age	Residual building age (year)	.00	44.00	25.2684	7.6405
Residual_age/S		.00	.70	.2575	.1307
Sunshine	Sunshine duration at the winter solstice (hour)	.00	8.00	3.5395	2.1504
Sunshine/S		.00	.12	3.328E-02	2.435E-02
Tree	Dummy of good greenery in the local district, if true, 1	.00	1.00	.3316	.4720
Shinjuku	Time distance from the nearest station to Shinjuku station (minute)	22.00	30.00	26.6263	3.4106
Station	Time distance to the nearest station (minute)	2.00	26.00	11.6158	4.9553
Road_width	Width of front road (m)	2.00	8.10	4.6358	1.2221

15.3 Hedonic Price Analysis

On the choice of appropriate functional form for hedonic price equations, it has usually been based on considerations for convenience in dealing with the problem at hand, rather than being specified on theoretical grounds (Halvorsen and Pollakowski, 1981). To include the effect on price, as well as unit price, for each factor, the following regression function was adopted:

$$P / S = \text{constant} + \sum_{i=1}^{k} c_i (X_i / S) + \sum_{i=1}^{k} d_i X_i + \varepsilon \qquad (15.1)$$

where, P: price of land lot and house, S: lot area, X_i: independent variables representing attribute i, c_i, d_i: parameters for $i \in \{1,\ldots,k\}$, and ε: error term.

Parameters c_i and d_i indicate the effects on price and unit price, respectively. Lot area was also entered as an independent variable, but its effect on P/S was not significant.

By regression analysis, we obtained a model with 16 independent variables. The first 11 variables were selected by a stepwise-incremental method on a significance level of F value at 0.05. The result of the regression is shown in Table 15.2. We strenuously included the last five variables, because bordering on to public green space, mixed land use, and the amount of trees appeared in the selection of regression on price through stepwise-regression analysis (Gao and Asami, 2001a), and these effects are of special interest. The significance levels of these five variables were satisfactory, too. This model explains up to 75.6 percent of the variation of the unit price. Validation tests showed that the 16 variables are statistically stable, and their estimates are reinforced by the findings of regression on price and logarithm regressions (Gao and Asami, 2001a).

The regression coefficients provided the estimates of the additional cost on the unit price for the marginal change of each variable. The hedonic prices

TABLE 15.2

Linear regression model on Unit Price

	Regression Coefficient (thousand yen/m² per unit)	Standard Coefficient	*t*-value	Significance
Constant	911.48		9.165	.000
Actual_FAR	127.56	0.182	3.215	.002
Station	−15.71	− 0.420	− 9.612	.000
Road_width	20.85	0.137	2.855	.005
Residual_age/S	568.62	0.401	6.419	.000
Landscape	−172.60	− 0.400	−8.463	.000
Shinjuku	−16.84	− 0.310	−6.596	.000
Frontage	5.80	0.113	2.383	.018
Good_pavement	42.00	0.114	2.798	.006
Parking_lot	38.17	0.153	3.536	.001
Build_quality	57.48	0.150	3.507	.001
Sunshine/S	947.61	0.125	2.669	.008
Greenery/S	21,454.70	0.268	3.138	.002
Greenery	−195.56	−0.257	−2.968	.003
Mixed_use/S	−17,476.55	−0.381	−2.438	.016
Mixed_use	238.41	0.404	2.635	.009
Tree	33.51	0.085	1.992	.048

of each attribute on price were then obtained by scaling the value with lot area (*S*). For instance, the regression coefficient of *actual_FAR* (actual building floor area/lot area) is 0.12756 million yen, hence the hedonic price of building floor area (= *actual_FAR* * *S*) equals 0.12756**S* million yen per 100 percent of the floor-area-ratio (FAR); likewise, the hedonic price of time distance to the nearest station (*station*) is -0.01571**S* million yen per minute.

Table 15.3 lists the hedonic prices for the significant attributes. Most of them are changing with lot area (*S*) (Parsons, 1990). Exceptions are sunshine duration (*sunshine*) and residual building age (*residual_age*), implying that they have almost equal influences, no matter how large a lot is. In addition, the analysis identified the critical values of lot size for the impacts of bordering on public green space (*greenery*) and large amounts of mixed land use (*mixed_use*). Bordering on green space increases price if the lot is smaller than 110 square meters; large amounts of mixed land use in the local district decreases price if the lot is smaller than 73 square meters. Public green space opens for anybody, which may have negative effect compared to having a garden of one's own. If a site is so narrow that having a garden is virtually impossible, adjacency to public green space may have a positive effect. If a site is sufficiently large to have its own garden, then no positive effect can be found for adjacency to public green space. Concerning the result of mixed land use, if the site area is small, the freedom to convert the land use is limited, and mixed land use only causes external diseconomy. As the site area gets large, the freedom for converting the land use increases, and the benefits of allowing mixed land use can be gained.

The hedonic price of sunshine duration is about 0.95 million yen per hour, suggesting that, all other conditions being the same, an additional hour of sunshine duration of a house results in a 0.95 million yen increase in the price of an asset. For the evaluation of local environments, this point is of particular importance. Whenever projects on lots exert external effects on

TABLE 15.3

Hedonic Prices of Attributes

Attribute	Hedonic Price
Building floor area (m²)	127.56 thousand yen/m²
Station (minute)	−15.71* *S* thousand yen/minute
Road_width (m)	20.85* *S* thousand yen/m
Residual_age (year)	568.62 thousand yen/year
Landscape (Y/N)	−172.60**S* thousand yen
Shinjuku (minute)	−16.84* *S* thousand yen/minute
Frontage (m)	5.80* *S* thousand yen/m
Good_pavement (Y/N)	42.00* *S* thousand yen
Parking_lot (car)	38.17* *S* thousand yen/car
Build_quality (Y/N)	57.48* *S* thousand yen
Sunshine (hour)	947.61 thousand yen/hour
Greenery (Y/N)	195.56* (109.7 − *S*) thousand yen
Mixed_use (Y/N)	238.41* (*S* − 73.3) thousand yen
Tree (Y/N)	33.51* *S* thousand yen

the surrounding land lots, sunshine duration is the main factor being affected. Based on this estimate, it becomes possible to value these effects.

Embedded in Equation 15.1, we also sought to identify, if any, nonlinear properties between sunshine duration and unit price. Sunshine duration in hours was replaced by dummy variables representing sunshine duration above a certain length. As the result of trials using two to six hours increased by a half hour as classifying points, the dummy variable indicating four hour or more sunshine was signified. This suggests that four hours could be regarded as a benchmark for appropriate sunshine in residential areas.

In the regression function, the effect of the landscape-dummy variable was identified to be negative. At first glance, the sign is contraintuitive. Nonetheless, within the designated landscape areas, building regulations often impose much stricter constricts on stories, building shape, setbacks, and so on, than other areas. The stricter restrictions might have reduced the willingness to pay for this attribute. In addition, the topography in landscape areas is mostly uneven. This may partly explain the strong negative effect of landscape dummy variable, too.

The amount of external effects is suggested by their hedonic prices induced from the regression coefficients, whereby we can evaluate local environment and develop new planning concepts. In the following, some applications of the presented results are described.

15.4 Analysis of Subdivision

Subdivision of land lots is known to be the main cause for worsening residential environment; it is still ongoing in Tokyo today. We analyzed this event with the linear-regression model. According to the model, when a lot is divided, the alteration in local attributes, including the changes of frontage, sunlight acquisition, the possible loss of parking space, and reduced space for planting, brings changes to the price of the lot. At the same time, the subdivision also exerts external effects on other lots.

Assume a block is composed of rectangular lots of 7 by 15 meters (Figure 15.1). Seven meters and 105 square meters are the medians of frontage and lot area in the formerly used sample. Small cross-marks can be found at the external walls of each house, and the figure beside the marks indicates the sunshine duration of that house, estimated with sunshine-duration software. One of the lots, lot A, is divided into a flag-shaped lot (sub 1) with frontage of 2.5 meters and lot area of 60 square meters, and a rectangular lot (sub 2) with frontage of 4.5 meters and lot area of 45 square meters. Furthermore, assume the floor-area ratio of both lots to be 1.00. In Table 15.4, the statistics before and after the subdivision are compared, using the hedonic prices given by Table 15.3. The subdivision usually sheds an external social effect to its surroundings, mainly by changing the sunshine acquisition. In this

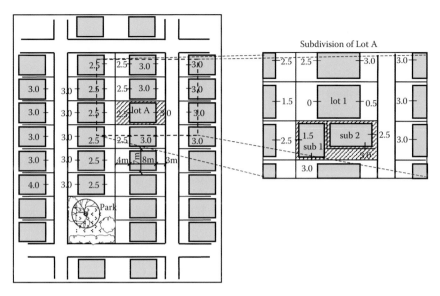

FIGURE 15.1

Subdivision of lots (unit of sunshine duration:hour).

TABLE 15.4

Effect of Subdivision of Lot A

	Factors of residential environment	Before	After		Hedonic price (million yen per unit)	Sum of hedonic prices (million yen)
			Sub1	Sub2		
Lot A	Actual_FAR	100%	100%	100%	0.12756*S	0
	Frontage (m)	7	2.5	4.5	0.0058*S	-2.22
	Parking_lot (car)	1	1	1	0.03817*S	0
	Sunshine (hour)	3	3	3	0.94761	2.84
	Aggregation					0.62
Lot 1	Sunshine (hour)	3	0.5		0.94761	-2.37
	Total					-1.75

case, the subdivision of lot A affects the lot to its north (lot 1) so this effect is included.

Table 15.4 shows that the subdivision increases the value of lot A by 0.62 million yen. However, by blocking the sunshine of lot 1, it produces a negative, external effect equal to 2.37 million yen, resulting in a total loss of 1.75 million yen. Theoretically speaking, the external effects caused by lot A should be covered by the owner of lot A, too. Thus, this example raised good evidence against the subdivision. By this analysis, the social effects are termed to money, so the result can be further adopted to deal with the equity problems among households.

The above example is quite representative. The result well agrees with the general criticisms against subdivisions. One may also note that, if subdivisions assemble in well-planted blocks, the loss will be aggravated by the additional negative effects caused by the loss of trees in those districts.

15.5 Benefits of Parks in Densely Built Residential Blocks

The provision of the parks is one of the main goals of the project to improve residential environments. Taking the land-readjustment project as an example, some land is allocated to parks and other public facilities, and the cost is balanced by the increased value of residential land due to a higher level of public facilities. In such processes, estimating the opportunity cost of parks is critical.

Consider the situation of turning a lot into a park in detached residential blocks. While the land provided for parks constitutes the cost, the park produces external effects on its surroundings that constitute the benefit. To simplify the analysis, we assume that the land in the middle of four lots is given over to a park, and there are only four lots adjacent to the park after the project. Three cases with different densities are assumed.

Case 1: Average lot size is 150 square meters

Case 2: Average lot size is 100 square meters

Case 3: Average lot size is 50 square meters

The original values of the assets and the external effects brought by the project are listed in Table 15.5. On computing the original values of the lots, we assume that time distance to the nearest station is 15 minutes; time distance to Shinjuku station is 22 minutes; the width of adjacent road is 4 meters; the frontage of the lots is 7 meters; and the residual building age of is 5 years. Besides, all characters of the local district are assumed standard so the values of the dummy variables indicating the characters of local district are 0.

TABLE 15.5

Effect of converting a Lot to a Park

	Hedonic price (million yen per unit)	Case 1:150m²		Case 2:100m²		Case 3:50m²	
		Before	After	Before	After	Before	After
Sunshine (hour)	0.94761	3	5	3	5	3	4
Number of lots adjacent to park	0.19556* (109.7 − S)	0	4	0	4	0	4
Social benefit (million yen)		−29.63		12.59		51.70	
Social cost (land provision) (million yen)		91.12		65.19		35.44	
Total effect (million yen)		−120.75		−52.60		16.26	

Table 15.5 reveals that when lot size is large, the external effects are negative; hence, it is difficult to balance the value of the original asset. However, the external effects of bordering on a park for small lots are strong enough to produce a positive opportunity cost. In case 3, where the area of the park is set to 50 square meters, the opportunity cost is 16.26 million yen. In a general meaning, this example suggests the feasibility of providing small parks, or the so-called pocket park, over large parks in densely built residential blocks.

15.6 Effects of Widening a Road

Years of practice in urban redevelopment reveals the hardness of the work in detached residential areas, especially when the density is high, where more individuals and interest groups are involved, and the relationships among them are complicated. Under this situation, controversies frequently arise concerning the treatment of small land lots and the equity issue between large and small land lots. Therefore, it is indispensable to study the performance of land lots with respect to lot size, and discussing the appropriateness of the policies in improvement projects on the basis of quantitative analyses is very important. We address the issue of lot size and configurations in urban detached residential areas in this section.

In the 1919–1950 Building Standard Law of Japan, 2.7 meters had been designated as the minimal legal width of residential roads. Although it was amended from 2.7 meters to 4 meters in 1950, a large number of narrow roads are left in urban areas. Roads narrower than 4 meters are now called "Item Two Road" and are termed to be the target of improvement. According to the Building Standard Law of Japan, Item 42(2), if not particularly designated, the space within 2 meters from the central line of roads should be regarded as road space, even though there are buildings along the road. The following analysis is made with an assumed road-widening project in the block shown in Figure 15.2.

The analysis is simplified: The road in Figure 15.2 is widened by w' on one side, and the lots on the other side are kept intact. The small section of the road framed by dashed lines in the figure is focused. There are two lots inside the section, which are assumed to have the same frontage (f) and depth (d) after the widening.

In this project, one lot gives a part of land to road. At the meantime, both lots get external benefits from the improvement, e.g., a widened road, better road pavement, and so on, so that the increased values of the lots will compensate for the land lost by these external effects. One may notify that landowners other than those of the two lots might also get social benefit from the improvement. However, this effect is difficult to estimate with the model developed in the former sections, so it is just neglected here.

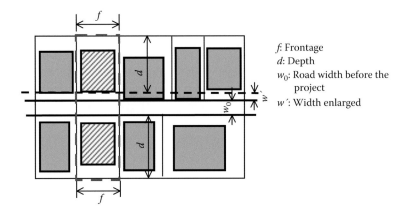

FIGURE 15.2
Assumed block for improvement project (*f*: frontage, *d*: depth, w_0: road width before the project, w': width enlarged).

In order to evaluate the social benefits and costs, the hedonic prices in Table 15.3 are used. First, some general assumptions are made:

Time distance to the nearest station (*station*): 5 minutes

Time distance to Shinjuku station (*Shinjuku*): 22 minutes

Assume that each lot has one parking lot. Like the former section, the characters of the local district are assumed standard, so the relevant dummy variables are zero.

We first discuss the net benefit of the improvement project when buildings inside the lots remain there. The social benefits and costs are represented by the changes of the value of the two lots (Gao and Asami, 2001b). Summing up these changes, we have the net benefit of the improvement in terms of the notations shown in Figure 15.2:

$$C_1 = -0.50062\ w'f - 0.0058\ w'f^2 + 0.0417\ w'\ df - 0.0417\ w'w_0\ f + 0.084\ df$$

(15.2)

By letting $C_1 \geq 0$, the marginal conditions of the net benefit can be derived. To concretize the problem, replace w_0 with 2.7 meters, w' with 1.3 meter, which means that the road is widened to 4 meters.

Next, the cost of the road is added into the consideration. This cost can be assumed a linear function of road area. Then the net benefit of the project is:

$$C_2 = C_1 + b(\ w_0 + w')f$$

(15.3)

where, *b* represents the per-square-meter cost of the road. The experienced value of road improvement for *b* (Setagaya Ward, Tokyo) is 0.04 million yen per square meter.

Then, suppose that the house on the widened side of road is reconstructed. With the reconstruction, the residual building age of the house is assumed to increase from 5 years to 30 years, and the newly built house recesses, because the borderline of the property sets back. As a result, the house in the north side gets an additional hour of sunshine. Furthermore, assume that the *FAR* of the lot increases from 100 percent to 150 percent.

The reconstruction cost of the house is also taken into account. The reconstruction cost is obtained with the user-survey data of the Housing Loan Corporation in 1999. A regression analysis with the 39,401 properties data reveals that the construction cost consists of a fixed cost and a linear term of floor area

$$\text{Construction cost} = 4.642 + 0.115 floor \text{ (million yen)} \quad (15.5)$$

where *floor* is the representation of the floor areas of houses, equal to $df \times FAR$. The R-square associated with the regression is 0.984.

Let the actual floor-area ratios of the lot before and after the project be indicated by FAR_b and FAR_a, respectively. The net benefit, C_3, can be calculated:

$$C_3 = C_2 + \underset{(Sunshine)}{0.94761} + \underset{(Residual\_age)}{0.56862 \times 25} + \underset{(Actual\_FAR)}{0.12756} \left[FAR_a \times df - FAR_b \times (d + w') f \right]$$

$$- (4.642 + 0.115 \times df \times FAR_a) \geq 0 \quad (15.6)$$

Then, the balancing curves of lot size before and after improvements can be derived.

In addition, for land lots whose size or depth is very small, the widening of road yields critical loss of land, thus the limitations on lots are indispensable. In practice, at least three constraints are necessary: $f \geq f_{min}$, $d \geq d_{min}$, and $S \geq S_{min}$, where, f_{min}, d_{min}, and S_{min} are the feasible lower boundaries of frontage, depth, and lot size, respectively. Their values ought to be defined in view of practical usage. Including these constraints, we have the marginal conditions and the circumscribed effective areas of lot size, which is shaded in Figure 15.3.

To get nonnegative social effects from the improvement project, the size and the corresponding frontage of land lots cannot go below the $C_3 = 0$ curves. For instance, when the frontage of lot is 12 m, the required minimum size is 152.6 m²; on the other hand, the maximum frontage for land lots with an area of 152.6 m² is 12 m.

With Equation 15.6 and the constraint functions, the frontage(s) associated with the vertices of effective areas in Figure 15.3 are easily computed:

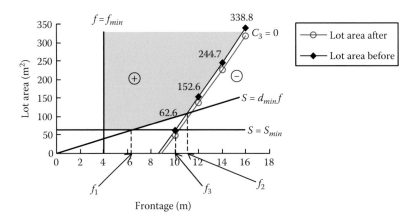

FIGURE 15.3
Effective area with building construction cost considered.

$$f_1 = S_{min}/d_{min}$$

$$f_2 = 66.313(-1.123 + \sqrt{0.317 + (1.123 - 0.0295d_{min})^2} + 0.0295d_{min}) \text{ if } f_2 > f_3$$

or

$$f_3 = 66.313(-1.123 + 0.0298\sqrt{1774.7 + S_{min}}) \text{ if } f_2 \leq f_3$$

Figure 15.3 reveals that a minimum lot size is necessary, and the limit on minimum lot size is increasing with respect to the frontage of lots. In addition, for a given-sized lot, the wider its frontage, the smaller the net benefit it gains from the project. When the frontage of lots is large, in other words, beyond max (f_2, f_3), the marginal curves of $C_3 = 0$ works; when the frontage is not as large, the limit is subject to the constraints on f_{min}, d_{min}, and S_{min}. For example, if $f_2 > f_3$, when the frontage of land lots is between f_{min} and f_1, the limit equals S_{min}, which is the minimal feasible size of land lot; when the frontage is between f_1 and f_2, it is determined by d_{min}, which is the slope of $S = d_{min}f$.

The above analysis implies that carefully considering the size of involved lots in redevelopment projects is very important. A minimal size level essentially exists, either constrained by the net benefits of the projects or by feasible usage of the lots.

15.7 Impact of Relaxing *FAR* Regulation

We are going to conclude the analysis with an exploration on the impact of relaxing *FAR* regulation in the improvement project. Although *actual_FAR* is

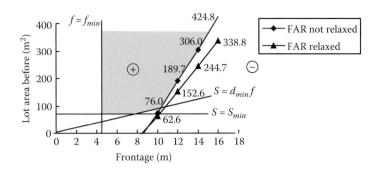

FIGURE 15.4
Effect of relaxing FAR regulation.

a value that landowners can freely determine, we assume that the owner of the land in Figure 15.2 decides to use the highest *FAR* permitted by city planning to build his house. This assumption helps us to estimate the marginal effect of relaxing *FAR*.

Since the analysis in the last part has studied the situation when FAR_b is 100 percent and FAR_a is 150 percent with Equation 15.6, a counterpart is created with the assumption that both FAR_b and FAR_a are 100 percent. The marginal conditions of the size with respect to the frontage of lots under the two cases are compared in Figure 15.4.

The curve with the *FAR*-relaxed case is lower than the without case, indicating that it is easier to get balance in the relaxed case. However, as lot size becomes smaller, the impact of relaxing *FAR* becomes smaller. In Figure 15.4, the intersections of the balancing curves and constraints on lot size and depth can be identified. If the frontage of lots drops below these thresholds, relaxing *FAR* regulation has little influence on reducing the limitations on lot size and depth. Furthermore, the interception of the balancing curves in horizontal-axis, another threshold value of frontage, is not affected by relaxing *FAR*. These results suggest that even though such policy is applied in redevelopment projects, land lots with very small frontage can hardly get benefits from the projects. Accordingly, it will not be a good incentive for the improvement project.

The above results are useful to clarify the specious concept that relaxing *FAR* regulation provides incentives for the redevelopment of densely built residential blocks. Asami (2000) ever argued that relaxing *FAR* regulation in these areas is risky, for it may lead to even higher densities, hence amplifing the difficulties of redevelopment. And, finally, the redevelopment of these areas might become public burdens. The analysis in this part suggests from another perspective that relaxing *FAR* might not be a proper solution for raising the incentives of redevelopment in densely built areas.

15.8 Conclusion

This chapter illustrated the application of GIS in the field of city planning and public policy. Using GIS, detailed spatial data were made possible for quantitative analyses, such as the hedonic price analyses, as shown above. As the results of hedonic price analyses with detail data in residential environments in Setagaya Ward in Tokyo, a price model explaining 75.6 percent of the unit price was derived, and local factors of residential environments were found significant among other explanatory variables. Since the residential environment of detached houses critically depends on local factors, the analysis result can be applied to the evaluations of a variety of physical planning and designs.

Specifically, the results of its application suggested the following:

1. External diseconomy due to the subdivision of a lot into small lots may be influential to surrounding lots.
2. Even though the value of a subdivided lot may increase, the overall value of the lots in the block may decrease.
3. Providing a pocket park in a densely built area may be socially beneficial, even taking its opportunity cost into account.
4. Application for the road-widening example suggested that the owners of smaller lots tend to have less incentive.
5. Lot size and the shape of lot influence the extent of the incentive.
6. Sometimes, the incentive cannot be efficiently increased through the relaxation of *FAR* regulation.

Local factors of residential environment play significant roles in the evaluation of hedonic price. This sort of quantitative analyses with detailed spatial data greatly help the judgment of urban policies. Since the fundamental tool supporting such analyses is the GIS-ready detailed spatial data, equipping such urban databases, as well as using such data for quantitative analyses, are essential for appropriate social judgment.

Acknowledgments

Valuable comments by Yoshihisa Asada, Tatsuo Hatta, Tomoo Inoue, Kiyohiko Nishimura, Atsuyuki Okabe, and Naohiro Yashiro were very useful in conducting this research. This research was partly funded by the Grant-in-aid for Scientific Research, the Special Coordination Funds for the promotion of Science and Technology from the Ministry of Education, Culture, Sports,

Science and Technology, and the research grant from the Land Institute of Japan. These are gratefully acknowledged.

References

Asami, Y., The rationale and negative effects of detached houses built on small land lots in urban areas, *Jutaku (Housing)*, 49(11), 5–8, 2000.

Blomquist, G. and Worley, L., Hedonic prices, demands for urban housing amenities, and benefit estimates, *J. Urb. Econ.*, 9, 212–221, 1981.

Boardman, A.E., Greenberg, D.H., Vining, A.R., and Weimer, D.L., *Cost-Benefit Analysis: Concepts and Practice, Prentice-Hall*, Upper Saddle River, New Jersey, 1996.

Cropper, M.L., Deck, L.B., Kishor, N., and McConnell, K.E., Valuing product attributes using single market data: a comparison of hedonic and discrete choice approaches, *Rev. Econ. Stat.*, 75, 225–232, 1993.

Gao, X. and Asami, Y., The external effects of local attributes on living environment in detached residential blocks, *Urb. Stud.*, 38, 487–505, 2001a.

Gao, X. and Asami, Y., Cost Benefit Analyses of Improvement Projects for Marginal Lot Size and Configurations, discussion paper No. 87, Department of Urban Engineering, University of Tokyo, 2001b.

Goodman, A.C., Hedonic prices, price indices and housing markets, *J. Urb. Econ.*, 5, 471–484, 1978.

Halvorsen, R. and Pollakowski, H.O., Choice of functional form for hedonic price equations, *J. Urb. Econ.*, 10, 37–49, 1981.

Hidano, N., Kameda, M., and Ando, S., Studies on the Benefit and cost of a setback regulation, proceedings of the Japanese Real Estate Institute, 14, pp. 125–128, 1998.

Huh, S. and Kwak, S., The choice of functional form and variables in the hedonic price model in Seoul, *Urb. Stud.*, 34, 989–998, 1997.

Japanese Institute of Architecture, Special Committee on Urban Housing Research, Basic terms of urban housing and housing land, Japanese Institute of Architecture, 1996.

Kameda, M. and Hidano, N., Basic study on the external environment of housing lots, proceedings of the Japanese Real Estate Institute, 13, 29–32, 1997.

Kain, J.F. and Quigley, J.M., Evaluating the quality of the residential environment, *Environ. Plann. A*, 2(1), 23–32, 1970.

Kanemoto, Y., Nakamura, R., and Yazawa, N., Using hedonic approach to estimate the value of environment, *Proceed. Environ. Sci.*, 2 (4), 251–266, 1989.

Kuwata, H., A study of building control for sunlight preservation in blocks, Papers on City Planning, 33, 787–792, 1998.

Li, M.M. and Brown, H.J., Micro-neighborhood externalities and hedonic housing prices, *Land Econ.*, 56, 125–141, 1980.

Mark, J.H. and Goldberg, M.A., A Study of the Impacts of Zoning on Housing Values Over Time, *J. Urb. Econ.*, 20, 257–273, 1986.

Mills, E.S. and Simenauer, R., New hedonic estimates of regional constant quality house prices, *J. Urb. Econ.*, 39, 209–215, 1996.

Parsons, G.R., Hedonic prices and public goods: an argument for weighting locational attributes in hedonic regressions by lot size, *J. Urb. Econ.*, 27, 308–321, 1990.

Pasha, H.A., Maximum lot size zoning in developing countries, *Urb. Stud.*, 29(7), 1173–1181, 1992.

Pogodzzinski, J.M. and Sass, T.R., Measuring the effects of municipal zoning regulations: a survey, *Urb. Stud.*, 28(4), 597–621, 1991.

Shilling, J.D., Sirmans, C.F., and Guidry, K.A., The impact of state land-use controls on residential land values, *J. Reg. Sci.*, 31, 83–92, 1991.

Tokyo Metropolitan Government, Land of Tokyo in 2002, Tokyo Metropolitan Government, Tokyo, 2003.

Yazawa, N. and Kanemoto, Y., Evaluation of residential environment through hedonic approach: application of GIS and reliability of predicted values, *Quarter. J. Hous. Land Econ.*, 36, 10–19, 2000.

16

Estimating Urban Agglomeration Economies for Japanese Metropolitan Areas: Is Tokyo Too Large?

Yoshitsugu Kanemoto, Toru Kitagawa, Hiroshi Saito, and Etsuro Shioji

CONTENTS

16.1 Introduction

Tokyo is Japan's largest city, with a population currently exceeding 30 million people. Congestion on commuter trains is almost unbearable, with the average time for commuters to reach downtown Tokyo (consisting of the three central wards of Chiyoda, Minato, and Chuo) being 71 minutes one-way in 1995. Based on these observations, many argue that Tokyo is too large and that drastic policy measures are called for to correct this imbalance. However, it is also true that the enormous concentration of business activities in downtown Tokyo has its advantages. The Japanese business style that relies heavily on face-to-face communication and the mutual trust that it fosters may be difficult to maintain if business activities are geographically decentralized. In this sense, Tokyo is only too large when deglomeration economies, such as longer commuting times and congestion externalities, exceed these agglomeration benefits.

In this chapter, we estimate the size of agglomeration economies using the Metropolitan Employment Area (MEA) data and apply the so-called Henry George Theorem to test whether Tokyo is too large. Kanemoto et al. (1996) was the first attempt to test optimal city size using the Henry George Theorem by estimating the Pigouvian subsidies and total land values for different metropolitan areas and comparing them. We adopt a similar approach, but make a number of improvements to the estimation technique and the data set employed. First, we change the definition of a metropolitan area from the Integrated Metropolitan Area (IMA) to the MEA proposed in Chapter 5. In brief, an IMA tends to include many rural areas, while an MEA conforms better to our intuitive understanding of metropolitan areas. Second, instead of using single-year, cross-section data for 1985, we use panel data for 1980 to 1995 and employed a variety of panel-data estimation techniques. Finally, the total land values for metropolitan areas are estimated from the prefectural data in the Annual Report on National Accounts.

16.2 Production Functions with Agglomeration Economies

Aggregate production functions for metropolitan areas are used to obtain the magnitudes of urban agglomeration economies. The aggregate production function is written as $Y = F(N, K, G)$, where N, K, G, and Y are the numbers of people employed, the amount of private capital, the amount of social-overhead capital, and the total production of a metropolitan area, respectively. We specify a simple Cobb-Douglas production function:

$$Y = AK^{\alpha}N^{\beta}G^{\gamma} \tag{16.1}$$

and estimate its logarithmic form, such that:

$$\ln(Y / N) = A_0 + a_1 \ln(K / N) + a_2 \ln N + a_3 \ln(G / N) \tag{16.2}$$

where Y, K, N, and G are respectively the total production, private-capital stock, employment, and social-overhead capital in an MEA. The relationships between the estimated parameters in Equation 16.2 and the coefficients in the Cobb-Douglas production function in Equation 16.1 are $\alpha = a_1$, $\beta = a_2 + 1 - a_1 - a_3$, and $\gamma = a_3$.

The aggregate-production function employed can be considered as a reduced form of either a Marshallian externality model or a new economic geography (NEG) model. The key difference between these two models is that the Marshallian externality model simply assumes that a firm receives external benefits from urban agglomeration in each city, while an NEG model posits that the product differentiation and scale economies of an individual firm yields agglomeration economies that work very much like externalities in a Marshallian model.

Let us illustrate the basic principle by presenting a simple example of a Marshallian model. Ignoring the social-overhead capital for a moment, we assume that all firms have the same production function, $f(n, k, N)$, where n and k are, respectively, labor and capital inputs, and external benefits are measured by total employment N. The total production in a metropolitan area is then $Y = mf(N/m, K/m, N)$, where m is the number of firms in a metropolitan area. Free entry of firms guarantees that the size of an individual firm is determined such that the production function of an individual firm $f(n, k, N)$ exhibits constant returns to scale with respect to n and k. The marginal benefit of Marshallian externality is then $mf_N(n, k, N)$. If a Pigouvian subsidy equaling this amount is given to each worker, this externality will be internalized, and the total Pigouvian subsidy in this city is then $PS = mf_N N$. If the aggregate-production function is of the Cobb-Douglas type, $Y = AK^\alpha N^\beta$, it is easy to prove that the total Pigouvian subsidy in a city is:

$$TPS = (\alpha + \beta - 1)Y \qquad (16.3)$$

The Henry George Theorem states that if city size is optimal, the total Pigouvian subsidy in Equation 16.3 equals the total differential urban rent in that city (see, for example, Kanemoto, 1980). Further, it is easy to show that the second-order condition for the optimum implies that the Pigouvian subsidy is smaller than the total differential rent if the city size exceeds the optimum. On this basis, we may conclude that a given city is too large if the total differential rent exceeds the total Pigouvian subsidy. The Henry George Theorem also holds in the NEG model, assuming heterogeneous products if the Pigouvian subsidy is similarly implemented. However, Abdel-Rahman and Fujita (1990) concluded that the Henry George Theorem is applicable even without the Pigouvian subsidy, although this result does not appear to be general.

Now let us introduce social-overhead capital, concerning which there are two key issues. The first of these concerns the degree of publicness. In the case of a pure, local public good, all residents in a city can consume jointly without suffering from congestion. However, in practice, most social-overhead capital does involve considerable congestion, and thus cannot be regarded as a pure, local public good. If the social-overhead capital were a pure, local public good, then applying an analysis similar to Kanemoto (1980) would show that the agglomeration benefit that must be equated with the total differential urban rent is the sum of the Pigouvian subsidy and the cost of the social-overhead capital. However, for impure, local public goods, the agglomeration benefit includes only part of the costs of the goods.

The second issue is whether firms pay for the services of social-overhead capital. In many cases, firms pay at least part of the costs of these services, including water supply, sewerage systems, and transportation. In the polar case, where the prices of such services equal the values of their marginal products, the zero-profit condition of free entry implies that the production function of an individual firm, $f(n, k, G, N)$, exhibits constant returns to scale with respect to the three inputs, n, k, and G, in equilibrium. In the other polar case, where firms do not pay for social-overhead capital, the production function is homogeneous of degree one, with respect to just two inputs, n and k.

Combining both the publicness and pricing issues, we consider two extreme cases. One is the case where the social-overhead capital is a private good and firms pay for it (the private-good case). In this case, the total Pigouvian subsidy is $TPS = (\alpha + \beta + \gamma - 1)Y = a_2Y$, and the Henry George Theorem implies $TDR = TPS$, where TDR is the total differential rent of a city. The other case assumes that the social-overhead capital is a pure, public good and firms do not pay its costs (the public-good case). The total Pigouvian subsidy is then $TPS = (\alpha + \beta - 1)Y = (a_2 - a_3)Y$, and the Henry George Theorem is $TDR = TPS + C(G)$, where $C(G)$ is the cost of the social-overhead capital. Although the evidence is anecdotal, most social-overhead capital adheres more closely to the private-good, rather than the public-good, case.

16.3 Cross-Section Estimates

Before applying panel-data estimation techniques to our data set, we first conduct cross-sectional estimation on a year-by-year basis. Table 16.1 shows the estimates of Equation 16.2 for each five year period from 1980 to 1995.

TABLE 16.1

Cross-Section Estimates of the MEA Production Function: All MEAs

Parameter	1980	1985	1990	1995
A_0	0.422**	0.440**	0.632***	0.718***
	(0.153)	(0.18)	(0.201)	(0.182)
a_1	0.404***	0.469***	0.528***	0.449***
	(0.031)	(0.039)	(0.043)	(0.037)
a_2	0.031***	0.026***	0.021**	0.020**
	(0.009)	(0.009)	(0.009)	(0.007)
a_3	0.015	-0.031	-0.124***	-0.086**
	(0.045)	(0.041)	(0.040)	(0.032)
$\bar{R}^2$	0.608	0.568	0.644	0.653

Note: Numbers in parentheses are standard errors. *** significant at 1 percent level; ** significant at 5 percent level.

The estimates of a_1 are significant and do not appear to change much over time. The estimates of a_2 are also significant, though they tend to become smaller over time. We are most interested in this coefficient, since $a_2 = \alpha + \beta + \gamma - 1$ measures the degree of increasing returns to scale in urban production. The coefficient for social-overhead capital, a_3, is negative or insignificant. As was observed and discussed in the earlier literature, including Iwamoto et al. (1996), this inconsistency implies the existence of a simultaneity problem between output and social-overhead capital, since infrastructure investment is more heavily allocated to low-income areas where productivity is low. Because of this tendency, less-productive cities have relatively more social-overhead capital, and the coefficient of social-overhead capital is biased downwardly in the Ordinary Least Squares (OLS) estimation. To control for this simultaneity bias, we use a Generalized Method of Moments (GMM) Three Stage Least Squares (3SLS) method in the next subsection.

The magnitudes of agglomeration economies may also be different between different size groups. Figure 16.1 shows estimates of the agglomeration economies coefficient a_2 for three size groups: large MEAs with 300,000 or more employed workers, medium-sized MEAs with 100–300,000 workers, and small MEAs with less than 100,000 workers, in addition to the coefficient for all MEAs. The coefficient is indeed larger for large MEAs, while for small and medium-sized MEAs, the coefficient is negative.

In addition to the simultaneity problem, OLS cannot account for any unobserved effects that represent any unmeasured heterogeneity that is cor-

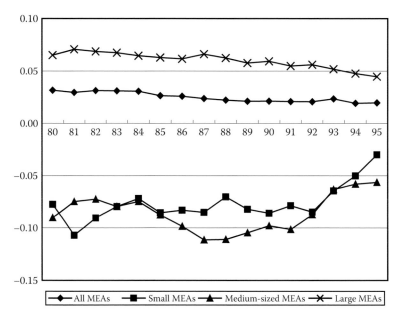

FIGURE 16.1
Movement of agglomeration economies coefficient a_2: 1980–95.

related with at least some of the explanatory variables. For example, the climatic conditions of a city that affect its aggregate productivity may be correlated with the number of workers because it influences their locational decisions. These unobserved effects also bias the OLS estimates. To improve these estimates, panel-data estimation with instrumental variables is used to eliminate biases caused by the simultaneity problem and any unobserved city-specific effects.

16.4 Panel Estimates

We first estimate the panel model whose error terms are composed of the city-specific, time-invariant term, c_i, and the error term, u_{it}, that varies over both city i and time t,

$$y_{it} = A_0 + a_1 k_{it} + a_2 n_{it} + a_3 g_{it} + b_t d_t + c_i + u_{it} \qquad (16.4)$$

where $y_{it} = \ln(Y_{it}/N_{it})$, $k_{it} = \ln(K_{it}/N_{it})$, $n_{it} = \ln(N_{it})$, $g_{it} = \ln(G_{it}/N_{it})$, and d_t is the time dummy. The use of the time dummy is equivalent to assuming fixed, time-specific effects. Table 16.2 shows fixed- and random-effects estimates. A random-effect model (RE) assumes the individual effects c_i are uncorrelated with all explanatory variables, while a fixed-effect model (FE) does not require the assumption. Though Hausman test statistics indicate the violation of the random-effect assumption for medium-sized MEAs and all MEAs, the estimation results of the random-effect model are more reasonable than those of fixed effects (see Wooldridge, 2002, Chapter 10, for Hausman test statistics).

TABLE 16.2

Panel Estimates

	All MEAs		Small MEAs		Medium MEAs		Large MEAs	
	FE	RE	FE	RE	FE	RE	FE	RE
a_1	0.279***	0.310***	0.354***	0.376***	0.281***	0.325***	0.170***	0.194***
	(0.015)	(0.014)	(0.030)	(0.027)	(0.021)	(0.020)	(0.029)	(0.026)
a_2	0.101***	0.031***	−0.016	−0.044	0.416***	0.096***	−0.044	0.059***
	(0.023)	(0.007)	(0.037)	(0.030)	(0.040)	(0.026)	(0.058)	(0.010)
a_3	−0.084***	−0.108***	−0.147***	−0.132***	0.145***	−0.061*	−0.151***	−0.113***
	(0.020)	(0.017)	(0.034)	(0.029)	(0.040)	(0.031)	(0.030)	(0.026)
$\bar{R}^2$	0.623	0.770	0.741	0.761	0.311	0.721	0.502	0.862
Hausman	39.6		11.3		132.5		21.3	
Chi (5 percent)	28.9		28.9		28.9		28.9	
Sample size	1888		528		896		464	

Note: Numbers in parentheses are standard errors. *** significant at 1 percent level; * significant at 10 percent level.

The random-effect estimates of the agglomeration coefficient a_2 are about 5 percent and 9 percent for large and medium-sized groups, but negative for small MEAs. Those for social-overhead capital are significantly negative for all groups.

The estimation results in Table 16.2 fail to eliminate the simultaneity bias, because both fixed- and random-effects models can deal only with the endogeneity problem stemming from the unobserved city-specific effects, c_i. Correlation between the random term, u_{it}, and social-overhead capital still provides a downward bias to the coefficients of social-overhead capital, and the results presented in Table 16.2 may reflect this problem. These considerations lead us to adopt a two-step GMM estimator, which, in this case, yields the 3SLS estimation in Wooldridge (2002, chap. 8, p. 194–8). (See Wooldridge, 2002, Chapter 8, pp. 188–199, and Baltagi, 2001, Chapter 8, for the explanation of GMM.)

We use time-variant instrumental variables (time dummies, k, n, and squares of n) and time-invariant ones (average snowfall days per year for the 30-year period 1971 to 2000 and their squares, and the logarithms of the number of preschool children and the number of employed workers who are university graduates in 1980). A major source of the bias could be the tendency of u_{it} to be negatively correlated with the social-overhead capital. Appropriate instruments are then those that are correlated with the social-overhead capital but do not shift the production function. The snowfall days per year satisfy the first property, because additional social-overhead investment is often necessary in regions with heavy snowfalls. It is not clear if the variable satisfies the second condition, since the inconvenience caused by snow may also reduce productivity. The logarithms of the numbers of preschool children and employed workers who are university graduates in 1980 are correlated with the regional-income level that negatively influences the interregional allocation of social-overhead capital. Since we use only the first year of our data set, they are exogenous for the subsequent production function, and it is reasonable to assume orthogonality with future idiosyncratic errors.

The revised estimation results are presented in Table 6.3. The coefficients of social-overhead capital are now positive but insignificant. The apparent simultaneity bias for social-overhead capital is only partially eliminated. Sargan's J and F values from the first regression, shown in Table 16.3, test the orthogonality condition for instrumental variables and the intensity of correlation between instruments and endogenous variables to be controlled (see Hayashi, 2000, Chapter 3, for Sargan's J statistics). While the F statistics are significant for all groups, the J statistics are significantly high for the two cases of all MEAs and medium-sized MEAs. The former results imply that the instruments we employed worked significantly well to predict the values of endogenous variables in the first regression. The latter results imply, however, that our instruments have failed to eliminate the simultaneity bias, at least in the two cases. The source of the bias is then likely to be the correlation between the instruments and city-specific, unobserved effects.

TABLE 16.3

GMM 3SLS Estimates

	All MEAs	Small MEAs	Medium MEAs	Large MEAs
a_1	0.518***	0.601***	0.479***	0.344***
	(0.030)	(0.066)	(0.047)	(0.048)
a_2	0.044***	0.027	0.053***	0.068***
	(0.005)	(0.018)	(0.013)	(0.007)
a_3	0.047	0.077	0.023	0.056
	(0.033)	(0.081)	(0.069)	(0.045)
J-statistics (D.F.)	16.28 (4)	5.73 (4)	24.57 (4)	3.78 (4)
Chi (5 percent)	9.49	9.49	9.49	9.49
1st stage *F*-statistics	216.85	81.10	105.19	91.73
Sample size	1888	528	896	464

Note: *** significant at 1 percent level.

One possible solution is to apply GMM estimation to time-differenced equations, as argued by Arellano and Bond (1991) and Blundell and Bond (1998). Both of these methods were tried but failed to yield satisfactory results. One cause of this failure is the fact that instruments that do not change over time cannot be used in the estimation of time-differenced equations.

The estimates of the agglomeration-economy parameter, a_2, are smallest for small MEAs and become larger for larger MEAs. An important difference from the OLS and RE estimates is that the sign of a_2 is positive even for small MEAs, which was negative in the earlier estimations. Accordingly, although our GMM 3SLS estimates display a number of shortcomings, they yield more reasonable estimates than those we have obtained elsewhere. We use the GMM 3SLS estimates as the relevant agglomeration-economy parameter in the next section.

16.5 A Test for Optimal City Sizes

Any policy discussion in economics must start with identification of the sources of market failure. In general, an optimally sized city balances urban agglomeration economies with diseconomy forces, and the first task is to check if these two forces involve significant market failure. On the side of agglomeration economies, a variety of microfoundations are possible, including Marshallian externality models (Duranton and Puga, 2003), new economic geography (NEG) models (Ottaviano and Thisse, 2003), and a reinterpretation of the nonmonocentric city models of Imai (1982) and Fujita and Ogawa (1982), as presented by Kanemoto (1990). Although the latter two do not include any technological externalities, the agglomeration economies that they produce involve similar forms of market failure. That is, urban agglomeration economies are external to each individual or firm, and a subsidy to increase agglomeration may improve resource allocation. This suggests that agglomeration economies are almost always accompanied by significant market failure.

In addition to these problems, the determination of city size involves market failure due to lumpiness in city formation. A city must be large enough to enjoy benefits of agglomeration, but it is difficult to create instantaneously a new city of a sufficiently large size, due to the problems of land assembly, constraints on the operation of large-scale land developers, and the insufficient fiscal autonomy of local governments. If we have too few cities, individual cities tend to be too large. In order to make individual cities closer to the optimum, a new city must be added. It may, of course, also be difficult to create a new city of a large enough size that can compete with the existing cities.

These types of market failure are concerned with two different "margins." The first type represents divergence between the social and private benefits of adding one extra person to a city, whereas the second type involves the benefits of adding another city to the economy. In order to test the first aspect, we have to estimate the sizes of external benefits and costs. To the authors' knowledge, no empirical work of this type exists concerning Japan. The Henry George Theorem can test the second aspect. According to this theorem, the optimal city size is achieved when the dual (shadow) values for agglomeration and deglomeration economies are equal. For example, the agglomeration forces are externalities among firms in a city, and the deglomeration forces are the commuting costs of workers who work at the center of the city, then the former is the Pigouvian subsidy associated with the agglomeration externalities, and the latter is the total differential urban rent.

Using the estimates of agglomeration economies obtained in the preceding section, we examine whether the cities in Japan (especially Tokyo) are too large. Our approach of applying the Henry George Theorem to test this hypothesis is basically the same as that in Kanemoto et al. (1996) and Kanemoto and Saito (1998). As noted in the preceding section, we consider two polar cases concerning the social-overhead capital. One is the case where the social-overhead capital is a private good, and firms pay for it. In this case, the total Pigouvian subsidy is $TPS = a_2 Y$, and the Henry George Theorem implies $TDR = TPS$, where TDR is the total differential rent of a city. The other case assumes that the social-overhead capital is a pure, public good, and firms do not pay its cost. The total Pigouvian subsidy is then $TPS = (a_2 - a_3)Y$, and the Henry George Theorem is $TDR - C(G) = TPS$, where $C(G)$ is the cost of the social-overhead capital.

Unfortunately, a direct test of the Henry George Theorem is empirically difficult, because good land-rent data is not readily available, and land prices have to be relied upon instead. Importantly, the conversion of land prices into land rents is bound to be inaccurate in Japan, where the price/rent ratio is extremely high and has fluctuated enormously in recent years. Roughly speaking, the relationship between land price and land rent is: Land Price

= Land Rent / (Interest Rate − Rate of Increase of Land Rent. In a rapidly growing economy, the denominator tends to be very small, and a small change in land rents generally results in a large change in land prices, as well as highly variable prices. For instance, the total real-land value of Japan tripled from 600 trillion yen in 1980 to about 1800 trillion yen in 1990, and then fell to some 1000 trillion yen in 2000. Given these possibly inflated and fluctuating land-price estimates and the inability to get good land-rent data, instead of testing the Henry George Theorem directly, we compute the ratio between the total land value and the total Pigouvian subsidy for each metropolitan area, to see if there is a significant difference in the ratio between cities at different levels of the urban hierarchy.

Our hypothesis is that cities form a hierarchical structure, where Tokyo is the only city at the top (see, for instance, Kanemoto, 1980; and Kanemoto et al., 1996). While equilibrium city sizes tend to be too large at each level of the hierarchy, divergence from the optimal size may differ across levels of hierarchy. At a low level of hierarchy, the divergence tends to be small, because it is relatively easy to add a new city. For example, moving the headquarters or a factory of a large corporation can easily result in a city of 20,000 people. In fact, the Tsukuba science city, created by moving national research laboratories and a university to a greenfields location, resulted in a population of more than 500,000. However, at a higher level, it becomes more difficult to create a new city, because larger agglomerations are generally more difficult to form. For example, the population-size difference between Osaka and Tokyo is close to 20,000,000, and making Osaka into another center of Japan would be arguably very difficult. We therefore test whether the divergence from the optimum is larger for larger cities, in particular if the ratio between the total land value (minus the value of the social-overhead capital when it is a pure, public good) and the total Pigouvian subsidy is significantly larger for Tokyo than for other cities.

The construction of the total land-value data for an MEA is as follows. The Annual Report on National Accounts contains the data on the value of land by prefecture. We allocate this prefecture data to MEAs, using the number of employed workers by place of residence. The first-round estimate is obtained by simple, proportional allotment. The problem with this estimate is that land value per worker is the same within a prefecture, regardless of city size. In order to incorporate the tendency that it is larger in a large city, we regress the total land value on city size, and use the estimated equation to modify the land-value estimates. The equation we estimate is: $\ln(V_i) = a \ln(N_i) + b$ where V_i is the first-round estimate of the total land value, N_i is the number of employed workers in a MEA, and a and b are estimated parameters. In the estimation, care has to be taken with sample choice, because in Japan, there are many small cities and very few large cities. If we include all MEAs, then the estimated parameters are influenced mostly by small cities. Since we are interested in the largest cities, we include the 19 largest MEAs in our sample. We drop the 20th largest MEA (Himeji), because it belongs to the same prefecture as the much larger Kobe, and the

TABLE 16.4

Total Land Values and Pigouvian Subsidies

MEA	Population	Land Value (a)	Pigouvian Subsidy 1 (b)	(a)/(b)	Social-Overhead Capital (c)	Pigouvian Subsidy 2 (d)	(a) – (c)/ (d)
Tokyo	30,938,445	518,810	9,493	55	133,310	1,613	239
Osaka	12,007,663	176,168	3,216	55	53,654	546	224
Nagoya	5,213,519	62,517	1,594	39	20,774	271	154
Kyoto	2,539,639	27,851	637	44	10.075	108	164
Kobe	2,218,986	21,913	575	38	12,345	98	98
Fukuoka	2,208,245	19,810	532	37	8,890	90	121
Sapporo	2,162,000	12,645	508	25	14,670	86	–23
Hiroshima	1,562,695	14,708	421	35	8,481	72	87
Sendai	1,492,610	12,529	377	33	7,604	64	77
Kitakyushu	1,428,266	11,059	311	36	6,719	53	82
Shizuoka	1,002,032	12,740	258	49	3,715	44	206
Kumamoto	982,326	6,505	206	32	4,892	35	46
Okayama	940,208	7,637	230	33	5,370	39	58
Niigata	936,750	7,519	231	33	5,698	39	46
Hamamatsu	912,642	11,489	242	47	3,707	41	189
Utsunomiya	859,178	8,021	223	36	3,551	38	118
Gifu	818,302	6,709	187	36	3,800	32	92
Himeji	741,089	6,143	205	30	4,640	35	43
Fukuyama	729,472	5,367	174	31	4,433	29	32
Kanazawa	723,866	7,412	182	41	3,957	31	112
Average		47,878	990	38	16,014	168	108

Note: Land value, Pigouvian subsidy, and social-overhead capital are in billion yen.

first-round estimate could then be seriously biased. The estimate of a is 1.20, with t-value 21.45. Assuming $V_i = AN_i^a$, we compute the total land value of an MEA by

$$V_i = \frac{\overline{V}}{\sum_j N_j^a} N_i^a . \tag{16.5}$$

Table 16.4 presents the total land value, the total Pigouvian subsidy, and social-overhead capital in the largest 20 MEAs in those cases where the production-function parameters are given by the GMM estimate for large MEAs in Table 16.3. The columns of "Pigouvian subsidy 1" and "Pigouvian subsidy 2" show the subsidies in the first case, $TPS = a_2 Y$, and the second case, $TPS = (a_2 - a_3)Y$, respectively. In both cases, the two largest MEAs, Tokyo and Osaka, have a significantly higher land value/Pigouvian subsidy ratio than the average city. This result supports the hypothesis that Tokyo is too large, but then it is likely that Osaka is also too large. These ratios are computed for the remaining years, and the same tendency exists. These results contrast with Kanemoto et al. (1996), who found that the land value/ Pigouvian subsidy ratio for Tokyo was slightly below the average for Japan's largest 17 metropolitan areas. One possible source of this difference is the land-value estimates used in this study, since the land values of Tokyo and Osaka are generally much higher than those in other Japanese cities.

Outside of Tokyo and Osaka, Shizuoka, Hamamatsu, and Kyoto also have high land value/Pigouvian subsidy ratios. This pattern is the same in the pure, public-good case. In the pure, public-good case, Sapporo has a negative ratio, because the value of social-overhead capital exceeds the total land value. This is caused by the fact that Sapporo is located on Hokkaido Island, and therefore receives a disproportionately high share of social-overhead capital investment.

16.6 Conclusion

Using the estimates of the magnitudes of agglomeration economies derived from aggregate production functions for metropolitan areas in Japan, we have examined the hypothesis that Tokyo is too large. In a simple, cross-section estimation of a metropolitan production function, the coefficient for social-overhead capital is either negative or statistically insignificant. The main reason for this is a simultaneity bias arising from social-overhead capital being more heavily allocated to low-income regions in Japan. In order to correct for this bias, we adopted panel-data methods. Simple fixed-effects and random-effects estimators still yield negative estimates. We also introduced instrumental variables and applied the GMM 3SLS to our panel data. The estimates become positive, but insignificant. The instrumental variables appear to reduce such bias, but they may not be strong enough to yield an unbiased estimate.

Using the GMM estimates for agglomeration economies, we also examined whether the Henry George Theorem for optimal city size is satisfied. Tokyo and Osaka have a higher land value/Pigouvian subsidy ratio than other cities. This indicates that Tokyo and Osaka are too large on the basis of this criterion. However, these results are tentative, and elaboration and extension in many different directions may be necessary.

Acknowledgment

This research is supported by the Grants-in-aid for Scientific Research No.10202202 and No. 1661002 of the Ministry of Education, Culture, Sports, Science and Technology.

References

Abdel-Rahman, H. and Fujita, M., Product variety, Marshallian externalities, and city sizes, *J. Reg. Sci.*, 30, 165–183, 1990.

Arellano, M. and Bond, S., Some tests of specification for panel data: Monte Carlo evidence and an application to employment equations, *Rev. Econ. Stud.*, 58, 277–297, 1991.

Baltagi, B.H., *Econometric Analysis of Panel Data*, 2nd ed., John Wiley & Sons, Chichester, 2001.

Blundell, R. and Bond, S., Initial conditions and moment restrictions in dynamic panel data models, *J. Econ.*, 87, 115–143, 1998.

Duranton, G. and Puga, D., Micro-foundations of urban agglomeration economies, in *Handbook of Urban and Regional Economics*, Vol. 4, Henderson, J.V. and Thisse, J.-F., Eds., North-Holland, Amsterdam, 2063–2117, 2004.

Hayashi, F., *Econometrics*, Princeton University Press, Princeton, New Jersey, 2000.

Iwamoto, Y., Ouchi, S., Takeshita, S., and Bessho, T., Shakai Shihon no Seisansei to Kokyo Toshi no Chiikikan Haibun (Productivity of social overhead capital and the regional allocation of public investment), *Financ. Rev.*, 41, 27–52, 1996 (in Japanese).

Kanemoto, Y., *Theories of Urban Externalities*, North-Holland, Amsterdam, 1980.

Kanemoto, Y., Optimal cities with indivisibility in production and interactions between firms, *J. Urb. Econ.*, 27, 46–59, 1990.

Kanemoto, Y., Ohkawara, T., and Suzuki, T., Agglomeration economies and a test for optimal city sizes in Japan, *J. Jpn. Int. Econ.*, 10, 379–398, 1996.

Kanemoto, Y. and Saito, H., Tokyo wa Kadai ka: Henry George Teiri ni Yoru Kensho (Is Tokyo too large? A test of the Henry George Theorem), *Hous. Land Econ. (Jutaku Tochi Keizai)*, 29, 9–17, 1998 (in Japanese).

Ottaviano, G. and Thisse, J.-F., Agglomeration and economic geography, in *Handbook of Urban and Regional Economics*, Vol. 4, Henderson, J.V. and Thisse, J.-F., Eds., North-Holland, Amsterdam, 2564–2608, 2004.

Wooldridge, J.M., *Econometric Analysis of Cross Section and Panel Data,*: MIT Press, Cambridge, Massachusetts, 2002.

17

Evaluation of School Redistricting by the School Family System

Yukio Sadahiro, Takashi Tominaga, and Saiko Sadahiro

CONTENTS

17.1 Introduction

Geographical Information System(GIS) is a set of tools for analyzing spatial objects and phenomena interactively in a computer environment. Since it treats geographical information, it is effective in educational administration to discuss geographical factors. For instance, we can visualize the location of schools, traffic networks, and public facilities as an integrated map. A map of schools and population distribution classified by ethnicity and race is useful for discussing the educational program desirable in each school. Calculating the average distance from home to school, we can evaluate a physical aspect of educational environment.

This paper aims to present potentials of GIS in educational administration. Potential applications of GIS in educational administration are threefold: analysis, planning, and evaluation. We discuss these subjects in turn in the following sections. We then show a methodology for treating school redistricting in GIS with a focus on the school-family system, a new concept in school cooperation. Applying the method to a concrete example of school redistricting in Tokyo, Japan, we will show the effectiveness of GIS in educational administration. In the last section, we summarize the conclusions.

17.2 Potential of GIS in Educational-Administration Research

17.2.1 GIS for Analysis in Educational-Administration Research

One typical usage of GIS in educational administration is spatial analysis of the present status of education in a region. Spatial analysis usually starts with visual analysis, which is followed by statistical and mathematical analysis. These steps are explained successively in the following.

Visual analysis is an initial examination of spatial phenomena in GIS (MacEachren and Taylor, 1994; Nielson et al., 1997; Slocum, 1998; Gahegan, 2000). Suppose, for instance, a map showing the location of schools, the number of students, and the population distribution of Hispanics (Figure 17.1a). A spatial variation exists in the number of students among school. Some schools have very few students, while others have so many students that they may be beyond their capacity. Since a strong correlation exists between the number of students and that of Hispanics, we suppose that a sudden increase of Hispanic students may have caused lack of schools, which has lowered the quality of educational environment.

If our interest lies in the regional variation in the grade of students, we may overlay a map of students with their grade on the map showing pop-

O   Elementary school

■ ■ ı   Major traffic road

FIGURE 17.1
GIS maps for visual analysis: (a) the number of students (white circles) and the distribution of Hispanics (gray shades); (b) traffic network (broken lines) and crime rate (gray shades).

ulation distribution classified by gender, age, ethnicity, and race. The quality of teachers and educational programs can also be visualized as attributes of schools to be discussed in relation to regional variation of students.

Visual analysis is also useful for assessing educational environment in a region. A map indicating the location of schools and individual students is useful for assessing the regional variance in the distance from home to school. Overlaying the maps of traffic networks and topography, we can evaluate the time distance instead of physical distance. Maps showing crime occurrences and land-use patterns may also work as indicators of educational environment (Figure 17.1b). Though we can do these analyses manually using paper maps, GIS drastically improves the efficiency and accuracy of analysis.

Visual analysis provides a lot of useful information about spatial phenomena. However, visual analysis is inherently subjective to some extent, because it primarily depends on the perception and evaluation of the analyst. Moreover, since the result is qualitative rather than quantitative, it often fails to lead persuasive result and discussion. Consequently, visual analysis is usually followed by statistical and mathematical analysis (Bailey and Gatrell, 1995; O'Sullivan and Unwin, 2002; Haining, 2003). Statistical analysis includes basic summary statistics, say, the number of students and teachers, the floor size of school buildings, the cost of maintaining educational resources, both human and physical, and so forth. They are calculated for individual schools, reported by the histogram, mean, variance, the maximum and minimum values, and often represented by the size of map symbols in GIS.

These descriptive measures are useful for generating research hypotheses. To test research hypotheses, statistical tests are performed. In addition to traditional statistics, spatial statistics are often used in GIS (Isaaks and Srivastava, 1989; Cressie, 1993; Diggle, 2003). Spatial statistics are a subfield of statistics focusing on the spatial distribution of stochastic phenomena. Whether or not the students of high grades are clustered in specific regions can be statistically tested at a given significance level. Spatial relationship between the location of schools and juvenile offenses can also be statistically evaluated.

Once a hypothesis is statistically supported, mathematical models are built to represent spatial phenomena. General spatial models include spatial-regression models (Bailey and Gatrell, 1995; Fotheringham et al., 2002), geo-statistics, spatial-point processes (Stoyan and Stoyan, 1994; Diggle, 2003), spatial econometrics (Anselin, 1988; Anselin and Florax, 1995), and spatial-choice models (Ben-Akiva and Lerman, 1985; Fischer et al., 1990; Smith and Sen, 1995). These models describe spatial phenomena in a formal manner using mathematical and statistical theories. Suppose, for instance, school choice of students, a kind of spatial choice behavior. Various factors are considered to affect school choice; the distance from home to school, the quality of education and facilities, the environment around school, and so forth. To measure the weight of each factor in school choice, a discrete choice model is often utilized. Collecting the data of school choice, we can estimate the model and evaluate quantitatively the degree of influence of each factor. This helps us in understanding one aspect of human behavior in education.

17.2.2 GIS for Planning in Educational-Administration Research

In educational administration, analysis is usually followed by plan making. For example, if the quality of education varies considerably among schools, a plan may have to be devised that assures all the schools of a certain quality of education. GIS can be utilized at this stage, that is, GIS supports decision-making in educational administration.

One effective tool for plan making is spatial optimization, which can be implemented in GIS (Drezner, 1996; Drezner and Hamacher, 2001). Spatial optimization is a collection of mathematical techniques that derives the spatial structure of variables optimal in a certain aspect. Imagine, for instance, location planning of elementary schools in a new town. There is no existing school, and financial status permits opening two elementary schools. For simplicity, we assume that the geography of the town is homogeneous and that the two schools provide the educational services of the same quality. In such a case, the distance from home to school is a critical factor in location planning; the average distance from home to school should be as small as possible. Spatial optimization gives a set of such locations that minimize the average distance from home to school.

Spatial optimization can consider not only a single element of educational environment, such as the distance from home to school, but also various factors simultaneously. In locating schools in a region, the traffic condition of the region, ethnic and racial balance among schools, the quality of teachers and programs also have to be taken into account. Administrative system is also an important element of educational administration. These factors are represented as variables, either qualitative or quantitative, and incorporated in mathematical calculations.

Besides facility location, spatial optimization includes network planning, location, and allocation of resources, shortest-path finding, and so forth. Network-planning techniques are useful for discussing the route of school buses, which is financially an important subject in educational administration. Location and allocation of educational resources, including teachers and facilities for education, can also be treated as a spatial-optimization problem.

Once spatial-optimization techniques are implemented in GIS, we can interactively compare alternatives for their decision-making (Lemberg and Smith, 1989; Ferland and Guénette, 1990; Armstrong et al., 1993; James, 1996). One may consider that the distance from home to school is very important and derive the optimal location of schools that minimizes the average distance from home to school. Others may think that the ethnic balance among schools is critical, which gives different optimal location of schools. If spatial-optimization techniques are implemented in GIS, we can try various viewpoints of a problem to be solved, derive their optimal solutions, and compare them using various measures.

17.2.3 GIS for Evaluation in Educational-Administration Research

After a program is executed, whether it works successfully is of great interest. To evaluate an educational program, we again use methods of spatial analysis with those of policy evaluation.

Take, for instance, the charter-school program. Unlike ordinary public schools, charter schools are run by nonformal organizations consisting of teachers, parents, and so forth, typically characterized by some unique educational programs. Charter schools don't have a certain district but overlap with those of ordinary schools, so that students can choose either a charter or an ordinary school. The main objective of the charter-school program is to provide students alternatives to ordinary schools, which leads to a competition among schools, and, consequently, improves the quality and efficiency of education. To evaluate the program, we need to know whether a charter school really draws students widely from its school district. Comparing the distributions of students of charter and ordinary schools using

GIS, we can easily examine whether the students of the charter school are distributed uniformly over its district, that is, whether the program works efficiently.

Similar to the charter-school program, the school-choice system also permits students to choose a school among several alternatives. However, the school-choice system aims not only to extend options for students but also to obtain a desirable balance in ethnicity and race in an educational environment, which is impossible in and ordinal school-district system. Consequently, whether the school-choice system is successful should be determined by the ethnic and racial balance achieved by the program. To this end, we compare the ethnic and racial compositions of individual schools and school districts using an overlay operation in GIS.

The school-bus system can also be evaluated efficiently in GIS. It not only is a safe transport system but also enables students to go to a school from distant places. This allows large school districts, and, consequently, reduction of schools for economic efficiency. To evaluate the school-bus system, we need to know whether students go to schools within a reasonable time. We can easily do this using network analysis in GIS if we have spatial data of traffic condition and bus routes.

17.3 GIS for School Redistricting

As discussed in the previous section, GIS has a great potential of contribution to educational-administration research. To illustrate this concretely, this paper presents an application of GIS to school redistricting. After general discussion in this section, an illustrative example is shown in the next section.

17.3.1 School Districting in Elementary and Lower-Secondary Education

Elementary and lower-secondary education is compulsory in many countries. To implement this, any student is assigned to one elementary and one secondary school. Though the school-choice system permits students to choose one school from several alternatives, such a program is exceptional.

Assignment of students to schools form a spatial structure called "school districts." In this paper, the term "school district" refers to the attendance area in which students are assigned to a certain school by the local school board, though "school district" often indicates the total area under the jurisdiction of the school board.

School districting is based on various factors (Campbell and Cunningham, 1990). Spatial factors include regional units, such as administrative units and census tracts, which reflect local communities. The distance to the school is also critical, especially in elementary education, where the school-bus system is not adopted. Young students should be assigned to nearby schools so that they do not have to walk so long to the schools. In some cases, traffic network is also considered; it is not desirable to cross roads of heavy traffic if students walk to school. As well as these spatial factors, nonspatial factors are also taken into account in school districting, say, school capacity and political social problems. Consequently, not all the school districts are based on administrative units; some units overlap with more than one school districts.

17.3.2 School Redistricting in Elementary and Lower-Secondary Education

In any country, school districts cannot be stable over time; they have to change, inherently corresponding to the distribution of students. In developing countries, for instance, new schools are built continuously with an increase of students. This always involves school redistricting. On the other hand, school redistricting accompanies school closures in developed countries, where students have been gradually decreasing. To keep economic efficiency, schools of few students have to be shut down. Old schools are often closed because school buildings need rebuilding.

School redistricting can occur without any change of schools. When a school-bus system is newly introduced, school districts are usually reexamined and changed. School redistricting is often involved in adoption of a new educational program, such as the charter-school and school-choice system.

Various factors have to be taken into account in school redistricting, as well as in school districting. In addition to spatial factors mentioned earlier, it is important to keep the racial and ethnic balance in each school. Variation in educational programs among schools should be considered, because every school develops its own program in response to the local demand for education. Consequently, school redistricting is a very complicated process of decision-making, which often takes considerable time.

17.3.3 School-Family System

School districts are usually determined separately for elementary and lower-secondary schools. Consequently, one district of an elementary school may overlap with districts of more than one lower-secondary school (Figure 17.2a). Students assigned to the same elementary school may go to different lower-secondary schools.

Recently, however, a new system called school-family system has been advocated (Los Angeles Annenberg Metropolitan Project, 2004). In this system, one lower-secondary school and several elementary schools form a *school family*. Schools in the same family cooperate with each other in the education of students. They share educational resources, such as teachers and school facilities, information of pupils and students, curriculum development, and teacher training. In the United States, a school-family system

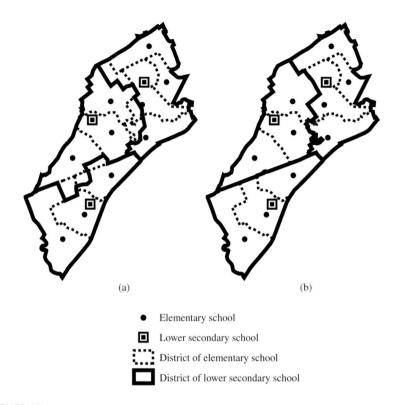

(a)   (b)

● Elementary school

▣ Lower secondary school

⌓ District of elementary school

▢ District of lower secondary school

FIGURE 17.2
Spatial structures of school districts: (a) conventional system; (b) school-family system.

is introduced to improve the quality of teachers and the economic efficiency of education.

In the school-family system, each elementary-school district is fully contained in the district of one specific lower-secondary school (Figure 17.2b). Consequently, in a spatial aspect, school districts of the two education levels show a completely hierarchical structure, in other words, a tree-like structure. All the students assigned to the same elementary school go to the same lower-secondary school. The school districts are determined simultaneously.

17.3.4 School Redistricting as a Spatial-Optimization Problem

One approach to a school-redistricting problem is spatial optimization using GIS (Garrison, 1959; Yeates, 1963; Heckman and Taylor, 1969; Bruno and Anderson, 1982; Greenleaf and Harrison, 1987; Schoepfe and Church, 1991; Lemberg and Church, 2000). In spatial optimization, objective function, variables, and constraints are used to formalize the problem to be solved.

An objective function is a quantitative measure of an alternative that we want to maximize or minimize. In elementary education, for instance, one option is the average distance from home to school that should be minimized in school redistricting. On the other hand, if the racial and ethnic balance is important, the variance in ratios of different races and ethnicities can be the objective function.

A variable is a value that affects the objective function. In school redistricting, variables represent the parameters in alternatives that can be manipulated in planners and policy makers, such as assignment of students to schools, openings and closures of schools, and so forth.

A constraint represents the condition that has to be satisfied in the optimal solution. In optimization of school districts, it often happens that too many students are assigned to one school, while others have only a few students. To avoid such a case, we can impose conditions to an optimization problem, such as the minimum and maximum numbers of students at each school. If some schools have to be closed to improve economic efficiency, we may limit the number of schools as a constraint. In such a case, schools to be closed and student assignments are discussed simultaneously, that is, both factors are included as variables in a spatial-optimization problem. Formalizing constraints as equations and inequalities, we solve a spatial-optimization problem to find values of the variables that minimize or maximize the objective function satisfying the constraints.

Figure 17.3 shows an example of a school-redistricting plan given by as a spatial optimization problem, where the average distance from home to school is the objective function that should be minimized in school redistricting. In this case, every student is assigned to the nearest school. Consequently, as shown in Figure 17.3a, school districts form a Voronoi diagram, which is a spatial tessellation where every location is assigned to its nearest generator point (Okabe et al., 2000). We can see that the new districts are considerably different from the present ones where not all the students are assigned to the nearest schools. If five schools are closed, school districts that minimize the average distance from home to school are given by Figure 17.3b. Each district is almost twice as large as that in Figure 17.3a.

17.4 School Redistricting in Kita Ward, Tokyo

This section presents an application of GIS to school redistricting in Japan, in order to illustrate how the problem is resolved in the GIS environment. The focus is on the physical environment of elementary education rather than its qualitative aspects, since GIS is effective, especially for discussing spatial factors.

In Japan, elementary and lower-secondary education is compulsory, and every student is assigned to one elementary and one secondary school. Assignment of students primarily depends on the administrative unit called chochomoku. The area and population of one chochomoku in urban areas range

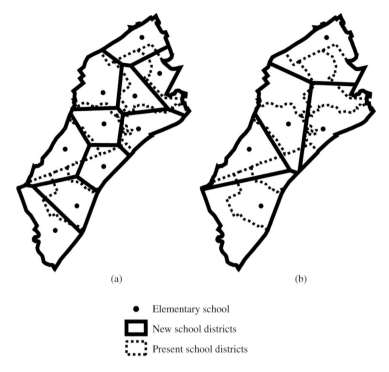

(a)   (b)

● Elementary school

▢ New school districts

⋯ Present school districts

FIGURE 17.3
Districts of elementary schools in Kita Ward, Tokyo.

from 0.1 to 0.5 and 500 to 5000, respectively, while a wider variation exists in rural areas. In principle, all the students in the same chochomoku are assigned to one nearby elementary school. Besides the distance to the school, traffic conditions are also considered. To avoid being involved in traffic accidents, students should not cross major-traffic roads in going to school. As well as these spatial factors, as described earlier, qualitative factors are also taken into account in determining school districts (for details, see Hayo, 1998).

17.4.1 Formulation of School-Redistricting Problem in Kita Ward, Tokyo

"Ward" is an administrative unit in Japan consisting of around 100 cho-chomokus. Kita Ward is located in the north of Tokyo. Kita Ward's area and population in 2004 were 20.59 km² and 316,000, respectively. Figure 17.4 shows the districts of elementary schools in Kita Ward. In this figure, we notice that schools are usually located near the centers of school districts. This implies that many students are assigned to their nearest schools, though there are some exceptions. Figure 17.5 shows the districts of lower-secondary schools in Kita Ward. Compared with Figure 17.4, schools are not always

FIGURE 17.4
Districts of elementary schools in Kita Ward, Tokyo.

located around the centers of school districts; more variation exists in the distance from home to school among students in the same district.

In Japan, both in urban and rural areas, students are gradually decreasing. For 2001, Kita Ward had 40 elementary schools for 11,609 students and 20 lower-secondary schools for 5710 students. At this time, schools are redundant in terms of the number of students per school. From 2001 to 2006, students are expected to decrease to 10,694 and 4470, respectively. To reduce the operational cost, school closures and redistricting are urgent topic for the local government of Kita Ward.

In school redistricting, two alternatives are examined. One follows the conventional system of school districts, that is, school districts of elementary schools and those of lower-secondary schools are determined separately. The other is the school-family system, where the school districts of two education levels are determined simultaneously so that every district of elementary school is completely contained in that of a lower-secondary school.

To discuss the above alternatives within the framework of spatial optimization, we consider two functions as the objective function in turn. One is the average distance from home to school. This assumes that distance is an important element of the physical environment of education, and that a

FIGURE 17.5
Districts of lower-secondary schools in Kita Ward, Tokyo.

shorter distance is better. The distance is actually critical in cities with high population density where students often walk to school rather than by school bus, and developing countries where the school-bus system cannot be adopted due to financial difficulties. The other objective function is the number of students whose assigned school changes by redistricting. The underlying opinion is that a drastic change of school assignment is not desirable in education, so it should be minimized. Whether the idea is reasonable depends on the case. It is widely supported, at least in Japan, that such a change should be avoided.

The variables in spatial optimization, which can be determined in educational administration, include school districts, that is, assignment of individual students to schools. In addition, schools to be closed are also considered, since students are decreasing in Kita Ward.

With respect to constraints, we compare two cases: One has no constraint, while the other assumes the school-family system, where one lower-secondary school and several elementary schools form a school family. The school-family system is formalized as a constraint where students of the same elementary school are assigned to the same lower-secondary school.

The above setting is summarized as follows:

Objective function:

Average distance from home to school

Number of students assigned to a different school by redistricting

Variables:

School districts

Schools to be continued and their districts

Constraints:

No constraint

School-family system

Since we have two objective functions, two variables, and two constraints, we solve eight spatial-optimization problems. Comparing two cases that share the same objective function and variables, one without constraints and the other with those of the school-family system, we can evaluate whether the school-family system is practically possible.

In this setting, the number of schools to be continued has to be given in advance. To this end we analyze how the number of schools affects the physical environment of elementary education. We formalize this as a spatial-optimization problem, where objective function is the average distance from students' houses to schools, and variables are schools to be continued and their districts. We successively solve the spatial-optimization problem changing the number of schools from five to 40 for elementary schools and from two to 20 for lower-secondary schools. In each problem, along with the derivation of the optimal solution of school closures and districts, we calculate the average and longest distances from home to school and the largest number of students of one school in order to evaluate the effect of limiting the number of schools.

Figure 17.6 and Figure 17.7 show the results for elementary and lower-secondary schools, respectively. At present, Kita Ward has 40 elementary schools and 20 lower-secondary schools; the average and longest distances from home to school are 367 and 1159 meters in elementary education, and 551 and 1731 meters in lower-secondary education, respectively. The distances reduce to 320 and 1045 meters in elementary education, and 488 and 1359 meters in lower-secondary education, respectively, if no school is closed. The distances and the largest number of students of one school monotonically increase with a decrease in the number of schools, as seen in the figures.

The largest numbers of students recommended by the Ministry of Education, Culture, Sports, Science and Technology are 720 (40 students, three classes, six grades) for elementary schools and 600 (40 students, five classes,

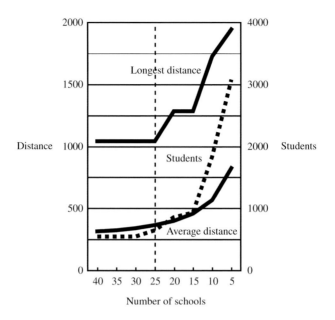

FIGURE 17.6

The number of schools and the physical environment of elementary education after optimization. The average and longest distances from home to school (bold lines), and the largest number of students of one school (broken lines). The figures on the left-hand axis indicate the average and longest distances, while those on the right-hand axis are the largest number of students of one school.

three grades) for lower-secondary schools, respectively (Ministry of Education, Culture, Sports, Science and Technology, 1956). Concerning elementary schools in Kita Ward, the largest number of students exceeds this standard when schools decrease from 25 to 20 (Figure 17.6). We thus determine that 25 elementary schools should be continued. Similarly, the minimum number of lower-secondary schools to be continued is set to 10, because the largest number of students exceeds the standard when schools decrease from 10 to eight (Figure 17.7).

17.4.2 School Redistricting Where the Average Distance from Home to School Is the Objective Function

This subsection adopts the average distance from home to school as the objective function.

We first discuss the case where no school is closed, and only school districts are changed so that the average distance is minimized. Two alternatives, one that introduces the school-family system and the other that follows the conventional system, are compared. The school-family system is further divided into two cases: in one case, one lower-secondary school and one to three elementary schools form a school family (school-family system I), while

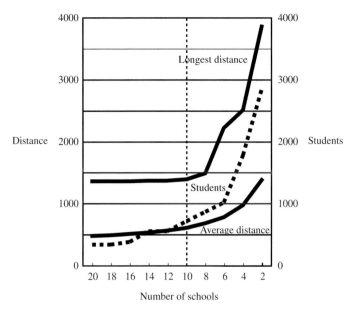

FIGURE 17.7
The number of schools and the physical environment of lower-secondary education after optimization. The average and longest distances from home to school (bold lines), and the largest number of students of one school (broken lines). The figures on the left-hand axis indicate the average and longest distances, while those on the right-hand axis are the largest number of students of one school.

in the other case, each lower-secondary school has two or three elementary schools (school-family system II). In general, school-family system II is adopted in urban areas. School-family system I, on the other hand, considers rural areas of low population density. Since districts of elementary schools are inevitably large, one lower-secondary school can be assigned to at most one elementary school to avoid students from going to very distant lower-secondary schools.

Table 17.1 shows the result of optimization. We first notice that optimization of school districts decreases the average and longest distances in both elementary and lower-secondary educations, independent of the introduction of the school-family system. On average, the distances decrease by 10 percent from those of the present. The table also shows that introducing the school-family system does not greatly affect the result of spatial optimization. Since the school-family system imposes an additional constraint in spatial optimization, the result is inevitably worse than that obtained for the conventional system. In elementary education, however, the school-family system does not change the longest distance from home to school at all, and only slightly raises the average distance from home to school. With respect to lower-secondary education, the longest distance from home to school increases from 1359 to 1603 meters. However, it is not a serious problem, because the distance is not so important in lower-secondary education as

TABLE 17.1

		Elementary school		
		Present status	After spatial optimization	
		Conventional system	School family system I	School family system II
Average distance from home to school	367m	320m (-12.8%)	327m (-10.9%)	342m (-6.8%)
Longest distance from home to school	1159m	1045m (-9.8%	1045m (9.8%)	1045m (-9.8%)
Ratio of students assigned to different school	NA	29%	29%	34%

		Lower secondary school		
		Present status	After spatial optimization	
		Conventional system	School family system I	School family system II
Average distance from home to school	551m	488m (-11.4%)	514m (-6.7%)	541m (-1.8%)
Longest distance from home to school	1731m	1359m (-21.5%)	1603m (-7.4%)	1603m (-7.4%)
Ratio of students assigned to different school	NA	38%	38%	39%

School family system I: One lower-seconday school has one to three elementary schools
School family system II: One lower-seconday school has two or three elementary schools

elementary education. Moreover, the longest distance is still shorter by 7 percent than that of the present.

We then consider the case of some of the elementary and lower-secondary schools that are closed. Setting 25 and 10 as the minimum numbers of the elementary and lower-secondary schools, we derive the school districts that are optimal in terms of the average distance from home to school. Since the average number of elementary schools to one lower-secondary school is 2.5, one lower-secondary school has two or three elementary schools in the school-family system (school-family system II).

Table 17.2 shows that optimization of school districts can partly compensate for the reduction of schools. The longest distance becomes 4 percent to 20 percent shorter in both elementary and lower-secondary educations, though the average distance increases by 12 percent to 17 percent in lower-secondary education. With respect to the school-family system, is consistent within that the school family system does not greatly affect the optimal physical environ-

TABLE 17.2

	Elementary school		
	Present status	**After spatial optimization**	
		Conventional system	**School family system II**
Average distance from home to school	367m	364m (0.8%)	376m (-6.8%)
Longest distance from home to school	1159m	1045m (9.8%)	1045m (-9.8%)
Ratio of students assighned to different school	NA	36%	38%

	Lower secondary school		
	Present status	**After spatial optimization**	
		Conventional system	**School family system II**
Average distance from home to school	551m	615m (+11.6%)	642m (+16.5%)
Longest distance from home to school	1731m	1397m (-19.3%)	1658m (-4.2%)
Ratio of students assighned to different school	NA	52%	52%

ment of education, especially at the elementary education level. From this we can say that the school-friendly system is a feasible option in school redistricting. The increase of the average distance from home to school is 3.3 (2.5 + 0.8) and 4.9 (16.5 – 11.6) percent in elementary and lower-secondary educations, respectively. The longest distance is unchanged in elementary education, while it increases by 15.1 percent in lower-secondary school. The students assigned to different schools are almost the same under conventional and school-family systems. In general, the school-family system does not drastically change the physical environment of education. From this we can say the system is a feasible option in school redistricting in Kita Ward.

17.4.3 School Redistricting Where the Number of Students Assigned to Different Schools Is the Objective Function

This subsection discusses the case where the number of students assigned to different schools is minimized by spatial optimization.

Table 17.3 shows the result when school redistricting is done without school closures. In this case, school assignment is given higher priority than

TABLE 17.3

	Elementary school			
		Present status	After spatial optimization	
		Conventional system	School family system I	School family system II
Average distance from home to school	367m	367m (0.0%)	375m (+2.2%)	494m (+34.6%)
Longest distance from home to school	1159m	1159m (0.0%	1159m (0.0%)	1727m (49.0%)
Ratio of students assighned to different school	NA	0%	8%	26%

	Lower secondary school			
		Present status	After spatial optimization	
		Conventional system	School family system I	School family system II
Average distance from home to school	551m	551m (0.0%)	553m (+0.4%)	606m (+10.0%)
Longest distance from home to school	1731m	1731m (0.0%)	1731m (0.0%)	2440m (+41.0%)
Ratio of students assighned to different school	NA	0%	3%	8%

	School family system I:	One lower-seconday school has one to three elementary schools
	School family system II:	One lower-seconday school has two or three elementary schools

the distance from home to school. Consequently, the average and longest distances are larger from those of the present, while students assigned to different schools are fewer than those in the previous section where the average distance from home to school is the objective function.

Both in elementary and lower-secondary education, school-family system I yields almost the same result as the conventional system. However, under school-family system II, where it is not permitted that one lower-secondary school has only one elementary school, the result is quite different; the average and longest distances from home to school drastically increase, as do the number of students being assigned to different schools.

We finally discuss the case where no school is closed and only school districts are changed so that the average distance is minimized. As seen in Table 17.4, the result is quite similar to that obtained when the average distance from home to school is minimized. In elementary education, the average distance slightly increases by introducing the school-family system.

TABLE 17.4

	Present status		
		After spatial optimization	
		Conventional system	**School family system II**
Average distance from home to school	367m	394m (-7.4%)	413m (+12.5%)
Longest distance from home to school	1159m	1075m (-7.2%)	1364m (+17.7%)
Ratio of students assigned to different school	NA	22%	27%

Lower secondary school

	Present status		
		After spatial optimization	
		Conventional system	**School family system II**
Average distance from home to school	551m	628m (+14.0%)	710m (+28.9%)
Longest distance from home to school	1731m	1731m (0.0%)	2171m (+25.4%)
Ratio of students assigned to different school	NA	47%	47%

The longest distance considerably increases in both education levels. The change of school assignment is almost the same if the school-family system is adopted.

The above results are consistent with those obtained in the previous subsection in that they are generally supportive of the school-family system. Under school-family system II, where one lower-secondary school has two or three elementary schools, introduction of the system yields a slight increase in the average distance from home to school and the number of students assigned to different schools. Though the longest distance increases, especially when schools are reduced to 25 and 10, the ratio is around 20 percent. Consequently, the school-family system where one family consists of one lower-secondary and two or three elementary schools is a feasible option in school redistricting in terms of physical environment of education.

17.5 Conclusion

In this paper, we have discussed potential applications of GIS in educational administration, taking the school-family system in school redistricting as an example. Spatial factors related to educational administration, typically the physical environment of education, can be represented as spatial data in GIS. Once spatial data are obtained, GIS works as an effective tool for discussing the spatial phenomena through a user-friendly, visual interface and mathematical functions. GIS helps us analyze spatial phenomena in an educational environment, devising an educational program, and evaluating it from various viewpoints.

Due to space limitations, we showed only one application of GIS in educational administration. We recommend that readers who interested in this subject refer to the books and papers listed at the end of this book. They greatly helped us to explore the potential of GIS in educational administration.

References

Anselin, L. and Florax, R.J G.M., *Spatial Econometrics: Methods and Models*, Kluwer, New York, 1998.

Anselin, L. and Florax, R.J G.M., *New Directions in Spatial Econometrics*, Springer, New York, 1995.

Armstrong, M.P., Lolonis, P., and Honey, R., A spatial decision support system for school redistricting, *URISA J.*, 5, 40–51, 1993.

Bailey, T.C. and Gatrell, A.C., *Interactive Spatial Data Analysis*, Taylor & Francis, London, 1995.

Ben-Akiva, M. and Lerman S., *Discrete Choice Analysis: Theory and Application to Travel Demand*, MIT Press, Cambridge, Massachusetts , 1985.

Bruno J.E. and Anderson, P.W., Analytical methods for planning educational facilities in an era of declining enrollments, *Socioecon. Plann .*, 16, 121–131, 1982.

Campbell, R.F. and Cunningham, L.L., *The Organization and Control of American Schools*, Prentice Hall, Englewood Cliffs, New Jersey , 1990.

Cressie, N., *Statistics for Spatial Data*, John Wiley, New York, 1993.

Diggle, P.J., *Statistical Analysis of Spatial Point Patterns*, Oxford University Press, Oxford, 2003.

Drezner, Z., *Facility Location: A Survey of Applications and Methods*, Springer, New York, 1996.

Drezner, Z. and Hamacher, W., *Facility Location: Applications and Theory*, Springer, New York, 2001.

Ferland, J.A. and Guénette, G., Decision support system for the school districting problem, *Op. Res.*, 38, 15–21, 1990.

Fischer, M.M., Nijkamp, P., and Papageorgiou, Y., *Spatial Choices and Processes*, North Holland, Amsterdam, 1990.

Fotheringham, A.S., Brunsdon, C., and Charlton, M., *Geographically Weighted Regression: The Analysis of Spatially Varying Relationships*, John Wiley, New York, 2002.

Gahegan, M., The case for inductive and visual techniques in the analysis of spatial data, *J. Geogr. Syst.*, 2, 77–83, 2000.

Garrison, W., Spatial structure of the economy, *Ann. Assoc. Am. Geogr.*, 49, 471–482, 1959.

Greenleaf, N.E. and Harrison, T.P., A mathematical programming approach to elementary school facility decisions, *Socioecon. Plann. Sci.*, 21, 395–401, 1987.

Haining, R., *Spatial Data Analysis: Theory and Practice*, Cambridge University Press, New York, 2003.

Hayo, M., *Districts of Elementary Schools*, Taga Shuppan, Kyoto, 1998 (in Japanese).

Heckman, L.B. and Taylor, H.M., School rezoning to achieve racial balance: a linear programming approach, *Socioecon. Plann .*, 3, 127–133, 1969.

Isaaks, E.H. and Srivastava, R.M., *An Introduction to Applied Geostatistics*, Oxford University Press, New York, 1989.

James, B., (1996). Use of Geographical Information Systems (GIS) mapping procedures to support educational policy analysis and school site management, *Int. J. Educ. Manage.*, 10, 24–31, 1996.

Lemberg, D.S. and Smith, E., Geographic Information Systems (GIS) and school facilities planning, *CASBO J.*, 54, 19–22, 1989.

Lemberg, D.S. and Church, R.L., The school boundary stability problem over time, *Socioecon. Plann .*, 34, 159–176, 2000.

Los Angeles Annenberg Metropolitan Project, www.laamp.org, 2004.

MacEachren, A.M. and Taylor, D.R.F., *Visualization in Modern Cartography*, Pergamon Press, Oxford, 1994.

Ministry of Education, Culture, Sports, Science and Technology, Closure policy of elementary and lower secondary schools, www.mext.go.jp/b_menu/shingi/12/chuuou/toushin/561101.htm, 1956.

Nielson, G.M., Hagen, H., and Mueller, H., *Scientific Visualization: Overviews, Methodologies, and Technologies*, IEEE Computer Society, Los Alamitos, New York, 1997.

Okabe, A., Boots, B., Sugihara, K., and Chiu, S.-N., *Spatial Tessellations: Concepts and Applications of Voronoi Diagrams*, John Wiley, Chichester, 2000.

O'Sullivan, D. and Unwin, D., *Geographic Information Analysis*, John Wiley, New York, 2002.

Schoepfe, O.B. and Church, R.L., (1991). A new network representation of a "classic" school districting problem, *Socioecon. Plann. Sci.*, 25, 189–197.

Slocum, T.A., *Thematic Cartography and Visualization*, Prentice Hall, Englewood Cliffs, New Jersey, 1998.

Smith, T.E. and Sen, A.K., *Gravity Models of Spatial Interaction Behavior*, Springer, New York, 1995.

Stoyan, D. and Stoyan, H., *Fractals, Random Shapes and Point Fields*, John Wiley, New York, 1994.

Yeates, M., Hinterland delimitation: a distance minimizing approach, *Prof. Geogr.*, 15, 7–10, 1963.

18

A Method for Visualizing the Landscapes of Old-Time Cities Using GIS

Eihan Shimizu and Takashi Fuse

CONTENTS

18.1 Introduction

Reproducing three-dimensional landscapes of old-time cities using contemporary maps and pictures brings precious information to historical studies. Since these documents often hold information retained by no other written or illustrated sources, the reproduced, three-dimensional landscapes complement previous life environments gathered from the archives.

Reproduced, three-dimensional landscapes are more than complements to recorded history. They provide researchers with a new tool for understanding the ancient life pattern of a city. In the reproduced conurbation, scholars are now able to virtually live next to the people of that time, and they can observe the ambience of that environment by going along the city streets using a walk-through simulation. Historians can experience past life space, and this experience enhances their vision of contemporary studies.

Although reproducing three-dimensional landscapes possibly provides a new tool for historical studies, this is not always an easy task, since old manuscripts are not usually geographically precise and are often distorted. The most difficult task in reproducing three-dimensional landscapes is to superimpose such distorted historical maps onto the precise maps of today.

It is important to point out here that research on *rubber sheeting*, that is, the geometric correction of maps for the conflation of maps from different source, has recently made much progress. Using this procedure, the geometric correction of historical maps should become possible, because it will make it easy to compare and overlay multiple maps from different time periods. It will, furthermore, give scales of distance, which are not generally a feature on old maps, and if we allow ourselves to ignore the changes in terrain, it will make possible the overlay of contour lines. We can, in consequence, bring the points of view of quantitative consideration and three-dimensional visualization into the analysis of ancient cartography. The value of historical maps as a database for research and information will thus be enhanced.

We have previously addressed applied research on such an incorporation of historical maps into Geographical Information System (GIS) (Shimizu, 2003). Rubber sheeting of some maps produced in the *Edo* and *Meiji* periods to the map-coordinate system currently used in Japan has already been done by using the piecewise geometric transformation based on the Triangulated Irregular Network (TIN) and planar affine transformation. (The *Edo* period ran between 1603 and 1868, and the *Meiji* period between 1868 and 1912. *Edo* was also the old name of Tokyo.) Since then, we have carried out applications, such as overlaying different maps from the *Edo* period to the present, quantitative analyses of land use, and visualization of the landscape of *Edo*. This chapter looks at the rubber-sheeting procedure and some situations in which the old maps are brought into use.

18.2 Rubber-Sheet Transformation of Historical Maps

The piecewise rubber sheeting based on TIN and planar affine transformation (White and Griffin, 1985; Saalfeld, 1985) has been very popular as a possible and effective map-conflation technique (Doytsher, 2000). We have applied this technique to the rubber sheeting of historical maps (Fuse et al., 1998; Shimizu et al., 1999; Shimizu and Fuse, 2003). More recently, its implementation has been reported by Niederoest (2002) and Balletti (2000).

The rubber-sheeting process that we used involved (Figure 18.1):

1. Identifying control points on a historical map and a modern one. The fixed geo-features that have remained stable were set as the

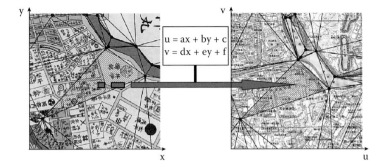

FIGURE 18.1
Piecewise rubber sheeting based on TIN and affine transformation.

    control points, for instance, temples, shrines, parts of castles, and similar substantial landmarks.

2. Forming a TIN over all the control points on both the historical and the modern map. Checking whether the relationship between both TINs is homeomorphic. If not, reforming one or the other TIN by hand, thus creating triangle pairs.

3. Performing a planar affine transformation for each triangle pair.

The advantages of this method are:

1. The topological relationship between features on the historical map is maintained, which is the most important property for the geometric correction of a map.

2. All control points are honored, that is, the control points on the historical map perfectly coincide with those on the present one. This is critical for an accurate comparison between the two maps.

Historical maps may not be correct in the geometric sense, being imprecise in the relative positioning of features. However, the linearity of objects, such as roads or a moat, may have been mapped with a fair degree of accuracy. If that is geographically correct, it is desirable to keep the "straightness" or other precision of representation of features on the rubber sheeting.

The affine transformation maintains linearity, and hence the integrity of alignment of features in a triangle is maintained. However, linear objects across adjacent triangles may be bent. Figure 18.2 shows *Tameike* in *Minato* Ward, Tokyo. The TIN produced is shown in Figure 18.2a. Figure 18.2b depicts a street that would have been straight in the *Edo* period but is now shown as bent.

To maintain the straightness of a street, we need to specify its ends as control points. However, this is not feasible, and the piecewise linear rubber-sheeting method cannot avoid this problem.

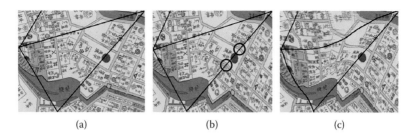

(a)   (b)   (c)

FIGURE 18.2
Piecewise linear and nonlinear transformation (a) original image, (b) affine transformation, and
(c) fifth-order polynomial transformation.

The piecewise, nonlinear, rubber-sheeting algorithm was developed by
Akima (1970, 1978). In place of the affine transformation, this method
employs the fifth-order polynomial transformation with conditions, such as
continuity and smoothness of adjacent piecewise transformations. This
method allows the avoidance of sharp, dogleg bends. However, the linear
features on the historical map are in general extremely distorted, as shown
in Figure 18.2c. We did not adopt this method, because this characteristic is
undesirable, since it destroys the primary or unique information within a
historical map.

18.3 Applications

18.3.1 Comparison of Maps from Different Times

We performed rubber sheeting for the following maps of different periods,
converting them to the plane rectangular-coordinate system of Tokyo:

  a. *Genroku-Edo-Zu* map (1693, *Edo* period)
  b. *Tenpou-Edo-Zu* map (1843, *Edo* period)
  c. *Jissoku-Tokyo-Zu* map (1892, *Meiji* period after *Edo* period)

Figure 18.3 shows the results, together with some of the Tokyo GIS data
in the areas surrounding *Tameike* in the *Minato* Ward, Tokyo. The Japanese
word *Tameike* means "reservoir" in English, and in the *Edo* period, there
actually was a large, natural reservoir, as shown in Figure 18.3a and Figure
18.3b. *Tameike* was filled in about 1880. The maps from different times virtually show us the history of urban development.

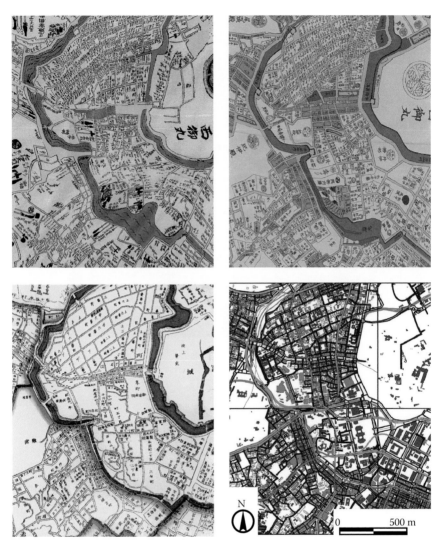

FIGURE 18.3
Historical maps from different periods (a) *Genroku-Edo-Zu* map (1693) (Source: Kochizu Shiryo Shuppan Publications), (b) *Tenpou-Edo-Zu* map (1843) (Source: Jinbunsha Publications), (c) *Jissoku-Tokyo-Zu* map (1892) (Source: Kochizu Shiryo Shuppan Publications), and (d) Tokyo GIS data.

18.3.2 Reproduction of a Digital-Elevation Model from a Historical Map

As the technique of representing landforms with contours was introduced into Japan in the early 1870s, there are no contour lines on the Japanese *Edo* period maps. The original Tokyo 1:5000 scale Survey Map produced by the Home Ministry (1888) is the oldest map on which contour lines were drawn.

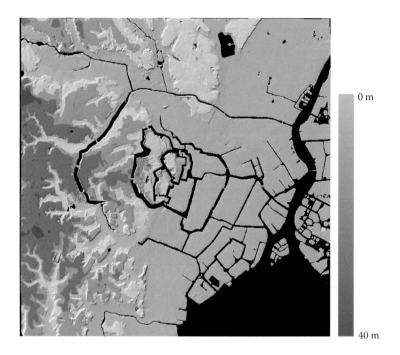

FIGURE 18.4
DEM of Tokyo in 1888.

This map contains 2-meter-interval contour lines and point data for elevation. We digitized these features and generated a Digital Elevation Model (DEM) for those times. Figure 18.4 shows the DEM with a 5-meter grid.

18.3.3 Analysis of the Relationship between Land Use and Topography

We overlaid the DEM on the historical maps to analyze the relationship between land use and topography. As an example, Figure 18.5 shows the geometrically corrected *Man'en-Edo-Zu* map (1860) of *Otowa* in *Bunkyo* Ward, Tokyo, overlaid by 5-meter-interval contour lines. With such a manipulation, we can consider the relationship between land use and topography during the *Edo* period. We can, for instance, see that the *Edo* government would have developed major streets by making use of the gentle slopes along the ridges and drainage lines. Figure 18.5 also shows that the *daimyos* (feudal lords) had their residences, which were large land lots north of the *Kanda-gawa* River, in a pleasant environment on a south-facing slope.

We can discern seven types of land occupation from the *Man'en-Edo-Zu* map: (1) *daimyo* (feudal lords), (2) *hatamoto* (direct retainers of the Shogun), (3) *kumi* (the lower class of *Samurai*), (4) *chonin* (commoners, such as retailers and artisans), (5) temples and shrines, (6) streets, and (7) rivers.

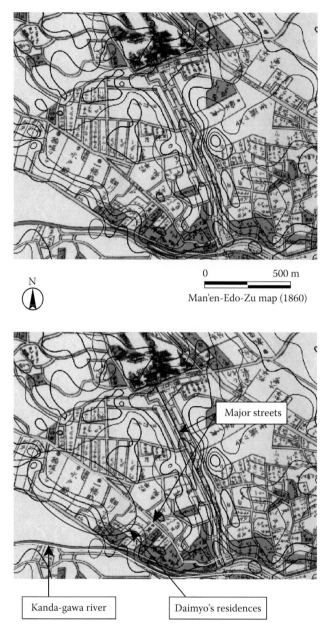

FIGURE 18.5
Overlay of contours on a historical map.

Figure 18.6 shows the land-use map of the same area as Figure 18.5 in 1860. This was derived by the hand digitizing of lot boundaries. Using this data, a quantitative analysis of the relationships between land use and elevation becomes much easier.

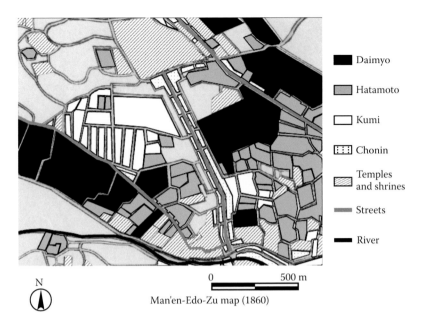

N

Man'en-Edo-Zu map (1860)

FIGURE 18.6
Land-occupation map in 1860.

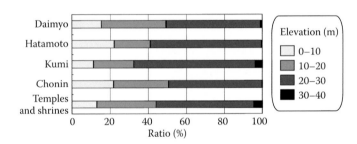

FIGURE 18.7
Land-occupation area: ratio by elevation.

We computed the area by elevation ratio for each type of land use over the whole *Bunkyo* Ward (Figure 18.7). We considered that the residences of those of higher rank, such as the properties of the *daimyo* and *hatamoto*, would have been located on higher land than the residences of those of lower-ranking *kumi* and *chonin*. However, as far as *Bunkyo* Ward is concerned, a striking difference was not apparent.

18.3.4 Reproduction of a Bird's-Eye View of Old Tokyo

We can give a bird's-eye view of old Tokyo by relating the historical maps to the DEM. Figure 18.8 — a bird's-eye view of *Edo* using the *Tenpou-Edo-*

FIGURE 18.8
(See color insert following page 176.) Bird's-eye view of *Edo* city (1843).

FIGURE 18.9
(See color insert following page 176.) Bird's-eye view of Tokyo in the *Meiji* period (1888).

Zu map (1843) — is such an example. At the end of the *Edo* period, *Edo* city was a huge, sprawling metropolis with more than 1 million inhabitants. The view gives a feeling of the extent and variation of the urban terrain. The area between the castle and the bay was reclaimed at the beginning of the period. The canals in the plain lying on the right side of the figure were developed for shipping and the reclamation of the waterfront.

Figure 18.9 gives a bird's-eye view of Tokyo in the *Meiji* period using the original Tokyo 1:5000 scale Survey Maps (1888). Each building is depicted as a basic three-dimensional model with its shade. This map is suitable for a fly-through animation of old Tokyo.

FIGURE 18.10
(See color insert following page 176.) Reproduction of the landscape depicted by Hiroshige's *Ukiyo-e* (a) *Nihonbashi-Yukibare* (Source: The Money Museum of UFJ Bank), (b) reproduction of the landscape, (c) *Tenpou-Edo-Zu* map (1843), and (d) current scene.

18.3.5 Reproduction of the Landscape of *Edo* City

Since we can study terrain data from historical maps, and we can read land use from them, if we prepare CG models that correspond to each land use, we can generally visualize the landscape of *Edo* from any viewpoint. Furthermore, the distant view of mountains can be introduced by integrating DEM data from the extensive general area, including the area covered by the historical map. The 50-meter DEM data produced by the Japanese Geographical Survey Institute are available for the whole country.

Ukiyo-e (wood block print) artists created many landscape prints of *Edo*. Among them, Ando Hiroshige (1797–1858), also known as Utagawa Hiroshige, is one of the most famous landscape artists. Hiroshige's work, along with that of the renowned Katsushika Hokusai, greatly influenced Western art.

We attempted to reproduce the landscape depicted by Hiroshige's prints. Figure 18.10a shows his famous print titled *Nihonbashi-Yukibare* (*Nihonbashi*, Clearing after Snow) from his series *Edo Meisho Hyakkei* (One Hundred Famous Views of *Edo*). The *Nihonbashi* ("Japan Bridge") area was the center

FIGURE 18.11
(See color insert following page 176.) Reproduction of a landscape not depicted by *Ukiyo-e* (1).

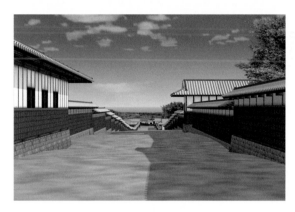

FIGURE 18.12
(See color insert following page 176). Reproduction of a landscape not depicted by *Ukiyo-e* (2).

of *Edo*. The fish market was located there, along with storehouses, and the place was a great symbol of wealth and plenty for the *Edo* Shogunate. Figure 18.10b is a reproduction of the landscape depicted by *Nihonbashi-Yukibare*. We can imagine that Hiroshige would surely have rearranged the relative positions of *Nihonbashi*, *Edo* Castle and *Mt. Fuji* in his artistic composition to permit the inclusion of these three famous views in/from *Edo* in a single printed sheet.

Our ultimate goal is to reproduce the landscape in *Edo* from any viewpoint and any direction and to explore the past concepts of city planning and urban development, which we may have forgotten in the modernization of Tokyo by placing too much emphasis on economic efficiency.

Figure 18.11 is a view from *Edobashi* (*Edo* Bridge) toward *Nihonbashi*. This view was not depicted by any *Ukiyo-e*. Although Tokyo Bay was also a famous view from *Edo*, we have been unable to find an *Ukiyo-e* that depicted this. Figure 18.12 shows a townscape and a view of Tokyo Bay from the current *Kasumigaseki,* the center of government. Needless to say, we no longer have a view of *Mt. Fuji* and Tokyo Bay.

18.4 Conclusion

The most remarkable characteristic of GIS is that it combines a wide variety of geographic data by relating them to a common coordinate system. By converting a historical map, through geometrical correction, to the current coordinate system using GIS, the value of the map in a data sense is much increased.

Following the conversion process, we can make an informed comparison between geographic features, such as land use portrayed on a historical map, and the modern scene depicted on current maps incorporating GIS. This is a fundamental task for historians using archival material. We also can clearly and impressively visualize the temporal and progressive change in the landscape from the past to the present, for instance, by employing techniques, such as morphing animation.

We can furthermore reproduce the old-time landscape from the historical map involving distant views of surrounding topography by introducing current digital-elevation data, as demonstrated in this chapter. The landscape thus reproduced is more real, being a scenario in which people from that time would have actually experienced in their daily lives, and from which they would presumably have derived comfort and pleasure. Compared with the reproduction of a landscape lacking distant views, it will provide more stimulus, suggestion, and implication for studies not only of the history of urban landscape but also of other domains, such as lifestyle, urban planning, and social development.

References

Akima, H., A new method of interpolation and smooth curve fitting based on local procedures, *J. Asoc. Comput. Mach.*, 17(4), 589–602, 1970.

Akima, H., A method of bivariate interpolation and smooth surface fitting for irregularly distributed data points, *ACM Trans. Math. Software*, 4(2), 148–159, 1978.

Balletti, C., Analytical and quantitative methods for the analysis of the geometrical content of historical cartography, *Int. Arch. Photogramm. Remote Sens.*, 33(B5), 30–37, 2000.

Doytsher, Y., A rubber sheeting algorithm for non-rectangular maps, *Comput. Geosci.*, 26, 1001–1010, 2000.

Fuse, T., Shimizu, E., and Morichi, S., A study on geometric correction of historical maps, *Int. Arch. Photogramm. Remote Sens.*, 32(5), 543–548, 1998.

Niederoest, J., Landscape as an historical object: 3D reconstruction and evaluation of a relief model from the 18th century, *Int. Arch. Photogramm. Remote Sens.,* and *Spatial Info. Sci.,* 34(5/W3), 2002.

Saalfeld, A., A fast rubber-sheeting transformation using simplicial coordinates, *Am. Cartograph.,* 12(2), 169–173, 1985.

Shimizu, E., Landscape visualization of old-time cities: focusing on Tokyo of the past, Sixteenth United Nations Regional Cartographic Conference for Asia and the Pacific, Okinawa, Japan, July 14–18, 2003, E/CONF.95/6/IP.8.

Shimizu, E. and Fuse, T., Rubber-sheeting of historical maps in GIS and its application to landscape visualization of old-time cities: focusing on Tokyo of the past, proceedings of the 8th International Conference on Computers in Urban Planning and Urban Management, 11A–3, 2003.

Shimizu, E., Fuse, T., and Shirai, K., Development of GIS integrated historical map analysis system, *Int. Arch. Photogramm. Remote Sens.,* 32(5–3W12), pp. 79–84, 1999.

White, M.S. and Griffin, P., (1985) Piecewise linear rubber-sheet map transformations, *Am. Cartograph.,* 12(2), pp. 123–131, 1985.

19

Visualization for Site Assessment

Hiroyuki Kohsaka and Tomoko Sekine

CONTENTS

19.1 Introduction

Numerous approaches for site assessment have been developed in geography to evaluate sites for housing and retail facilities (Orford, 1999; Jones and Simmons, 1990). These approaches evaluate a site in terms of two factors, such as the site itself and its location. Site factor is related to the lot in which a facility may be located and the physical environment directly related to the facility. Location factor is connected with its surrounding, which provides opportunity of use or demand. Recently, site-assessment approaches have been performed on GIS to handle very complicated consumer markets (Birkin et al., 2004).

Accessibility is one of the major elements for the location factor in site assessment. Accessibility is measured from two sides, demand and supply. The measure of accessibility from the residential site to retail and service facilities is related to evaluate a housing site from the demand side. Five types of accessibility measures have been proposed: 1) container index, 2) minimum distance, 3) cumulative opportunity, 4) gravity potential, and 5) space–time (Kwan 1998). Talen and Anselin (1998) point out that the choice of accessibility measure has to be considered very carefully using a case study of the geographic-accessibility measures to public playgrounds at the census-tract level.

For site assessment to a retail or service facility, it is necessary to evaluate whether a site will be able to attract a certain volume of sales. Evaluation methods have been developed, such as 1) rating model, 2) regression model, and 3) spatial-interaction model (Birkin et al., 2002). The first is related to compare relative scores for sites, and the second and third can predict the sales volume using mathematical models. Accessibility for supply side is measured in site assessment for retail and service facilities. In the rating model, buffer technique is used to determine a straightforward, "accessible" area followed by overlay technique to clip out this buffer area. However this buffer/overlay approach has some shortcomings, in the point that transport network, natural or man-made barriers, and competition with already established outlets are not taken into account (Geertman, et al., 2004). However, this approach is widely used in practical site assessment by the reason of its simplicity (for example, site assessment for petrol forecourts is referred to in Birkin et al., 2003).

When these site-assessment approaches are applied to practical scenes, many problems have been pointed out. One of the critical issues is the accuracy of analytical results. Inaccurate results cannot be guaranteed to clear the hurdle of a resident's satisfaction or a client's sales target. The reliability of the end result to reduce the risk of a wrong or misleading decision is important to site assessment (Van der Wel, et al., 1994). The decision-maker therefore wants to reveal the extent to which uncertainty affects the "decision space."

The presentation of uncertain information is one use of visualization in the GIS community. The extra visual attribute that a visualization environment provides can be used to add a further dimension to a map, in order to judge "truth" on GIS by measures, of uncertainty, error (accuracy), variation, validity, reliability, stability, or probability (MacEachren, 1995). The visualization techniques to display uncertainty include side-by-side, overlay, and merged displays (Beard and Buttenfield, 1999). The merged display makes use of a bivariate map as a representation of quantitative data and reliability of those data.

For example, visualization techniques are applied to convey classification of uncertainty in classified imagery and soil maps (Fisher, 1994a; 1994b). For classified imagery, the uncertainty inherent in the assignment of a pixel to a class is conveyed by making the value or color of a pixel proportionate to the strength of it belonging to a particular class. Gahegan (2000) depicts a false color satellite-image fragment of an agricultural area, where vertical offset is used to represent the probability (as determined by a classifier) of a pixel being classified as "wheat."

This paper tackles improving the accuracy of site assessment using suitable visualization techniques to reduce the risk of a misleading site selection. In the second section, the visualization is applied to display classification uncertainty in an accessibility map to ophthalmic clinics. The third section performs a highly accurate simulation as a site-assessment approach for a car dealer to reveal "truth" as an inaccessible site. The last section discusses a mechanism to judge whether a highly accurate approach should be applied in the practical scenes.

19.2 Multilevel Measures of Accessibility and Its Spatial Variation within Residential Districts

19.2.1 Accessibility Measured at the Residential-District Level

As a case study in this section, accessibility is measured from residential districts (Cyocyo-aza) to ophthalmic clinics in Matsudo City, Chiba Prefecture, Japan. Matsudo is one of the satellite cities in the Tokyo metropolitan area. Its area is 61 square kilometers, and 19 ophthalmic clinics are located within the city. The shortest-path distance to the nearest clinic is measured using the second method in five accessibility measures mentioned above. Figure 19.1 shows location (+) of clinics and centroids of residential districts (residential point;▲) on the road network of the northwest part of Matsudo City. The shortest-path distance from each residential point to the nearest clinic is measured on the actual road network using network analysis of ArcView (Sekine, 2003).

Figure 19.2 shows statistical distribution of the shortest-path distance for 343 residential districts. The average of the distance is 1177 m, and its standard deviation is 575 m. By considering such a distribution of the distance, the degree of accessibility is divided into four accessibility levels, as follows: "good" is shorter than 750 m, "normal" is 750 m to 1500 m, "bad" is 1500 m to 2250 m, and "very bad" is longer than 2250 m. All residential districts in Matsudo City are classified into four levels of accessibility in Figure 19.3.

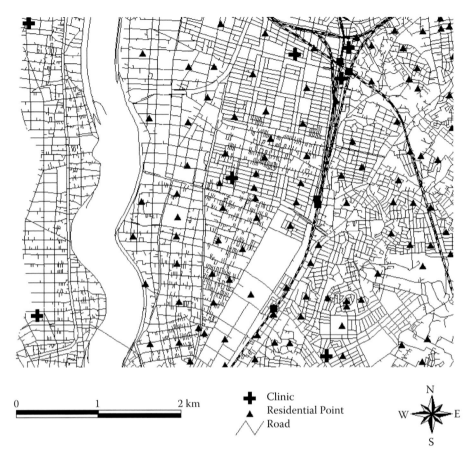

0   1   2 km   ✚ Clinic
           ▲ Residential Point
           ⋀⋁ Road

FIGURE 19.1
Location of ophthalmic clinics and centroids of residential districts on road network.

According to this result, residential district "A" shown in Figure 19.3 was assessed as "good" in terms of the accessibility to ophthalmic clinics.

19.2.2 Accessibility Measured at 100 M Mesh Level

Now let us measure the accessibility at finer level. The 1-kilometer mesh constructed in the Basic Area Mesh System[1] is divided into 10 equal segments for each side to create 100 m mesh. The shortest-path distance to the clinics is measured from the centroids of 6089 100 m meshes constituting Matsudo City. Figure 19.4 shows the four accessibility levels at 100 m mesh level.

The residential district A consists of 22 meshes, as shown in Figure 19.5. Fourteen meshes are assessed as "good" accessibility, and eight meshes are assessed as "normal." Therefore, we can recognize variation in accessibility

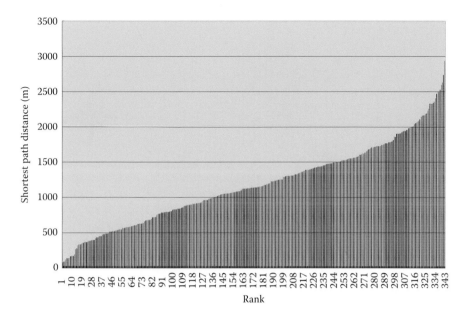

FIGURE 19.2
Shortest-path distance to the nearest ophthalmic clinics from residential points.

within this district. An issue may be raised by the residents in the meshes assessed as "normal," because they will find that their residential place is "normal" in spite of being assessed as "good" at the district level. This is known as modifiable area-unit problem (MAUP) and is particularly important for the residents in the case of lowering the accessibility level. In this case, the site assessment gives wrong information to them.

To examine such a variation in accessibility within all residential districts of Matsudo City, the accessibility map at the district level (Figure 19.3) was intersected with one at 100 m mesh level (Figure 19.4). Table 19.1 shows the variation volume of accessibility between two levels. The diagonal cells represent no change in accessibility level. These ratios are about 70 percent for "normal," and about 60 percent for "good," "bad," and "very bad." The ratios to the lower accessibility level are 33 percent for "good" and 15 percent for "normal." Inversely, the ratios to raise the level are 13 percent for "normal," 28 percent for "bad," and 36 percent for "very bad." For good or bad, it became clear that 30 percent to 40 percent of meshes have different accessibility levels from the one measured at the district level. The degradation of accessibility level, in other words, the rate at which spatial analysis at the district level over-assesses, amounts to 15 percent to 33 percent. And more than 30 percent of meshes within the district assessed as "very bad" raise the accessibility level. The result of this analysis means that the accessibility measured at the district level has not enough accuracy in practice.

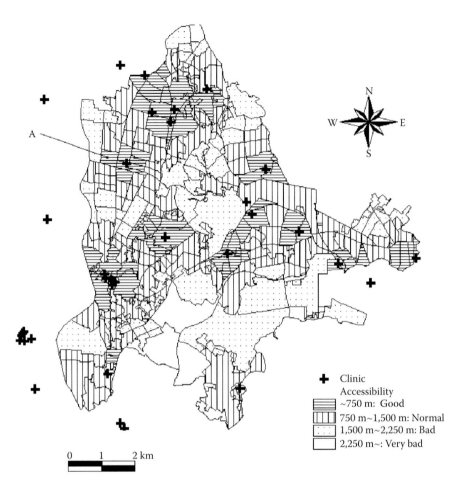

FIGURE 19.3
Accessibility to ophthalmic clinics at district level.

19.2.3 Visualization of Spatial Variation in Accessibility within a Residential District

Two stages of the selection process will be adopted in the selection of a residential site within a city. The first stage selects a residential district in the city, and the second selects a site within the district. Therefore, it is necessary to position the accessibility at an individual site or at the 100 m mesh level in the range of accessibility for the whole city, as shown at the residential-district level. To represent accessibility at the residential-district level while holding enough accuracy, three visualization techniques are proposed in the following.

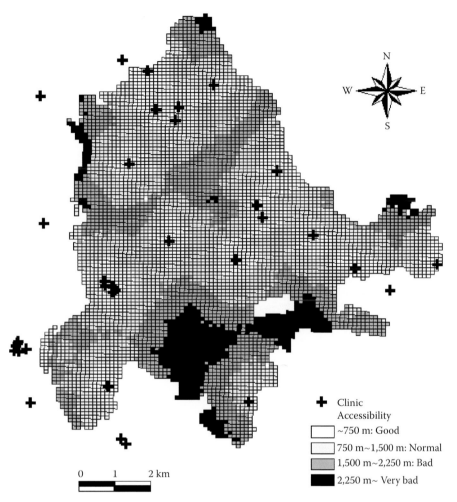

FIGURE 19.4
Accessibility to ophthalmic clinics at 100 m mesh level.

19.2.3.1 Bivariate Map of Accessibility and Its Variability

The first visualization is the overlay display in which accessibility and its variability is simultaneously represented as a bivariate map. Figure 19.6 is a bivariate map in which accessibility is classified into four levels, and its variability within the district is classified into four levels, such as 0 percent, 1 percent to 25 percent, 26 percent to 50 percent, and 51 percent or more. If the variability is zero, then accessibility is distributed uniformly within the district. If the variability is 51 percent or more, it means half or more of meshes consisting of the district differ from the accessibility level assessed at the district level.

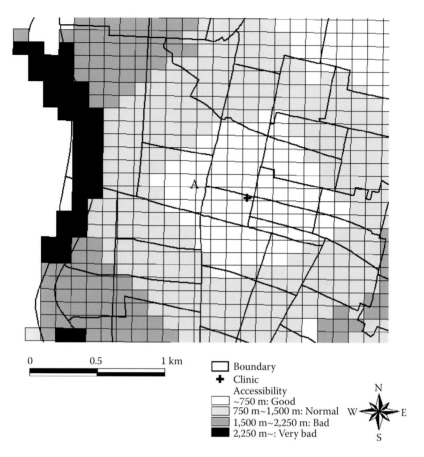

FIGURE 19.5
Spatial variation in accessibility within residential district A.

TABLE 19.1

District level	100m mesh level			
	Good	Normal	Bad	Very bad
Good	1521(67.6)	555(24.7)	125(5.5)	50(2.2)
Normal	575(13.4)	3051(71.2)	630(14.7)	30(0.7)
Bad	76(2.8)	688(24.9)	1683(61.0)	313(11.3)
Very bad	0(0.0)	14(2.4)	195(33.4)	375(64.2)

   This map shows that classification accuracy (uncertainty) is different even among the districts assessed as "good" accessibility (see district A, which is "good" in accessibility and is 26 percent to 50 percent in variability). Therefore, we can avoid making a misleading decision in evaluating a residential district using this map. However, this visualization shows the level of accu-

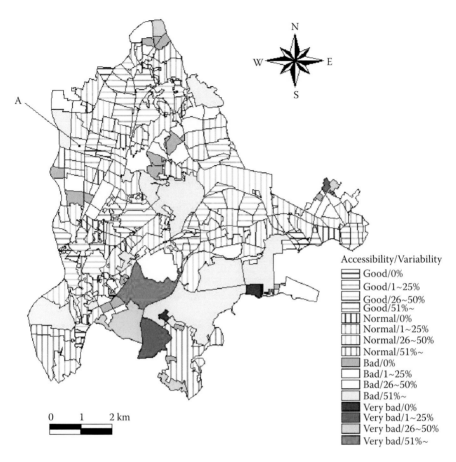

FIGURE 19.6
(See color insert following page 176.) Bivariate map of accessibility at district level and its internal variability.

racy for the districts, but cannot show where and to what degree accessibility differs inside the district.

19.2.3.2 Composite Map of Accessibility by Two-Level Visualization

The second is two-level visualization technique. Usually, a map is constructed at one level of spatial resolution. However, there may be a transitional zone in which the accessibility will be changed from one level to another level. For the district including such a zone, the result of accessibility should be represented at more detail (high) spatial resolution. Two-level visualization will be used for such a situation to hold accuracy of the result.

Now, let us apply two-level visualization to the accessibility in Matsudo City. If the accessibility level for a residential district is the same level for

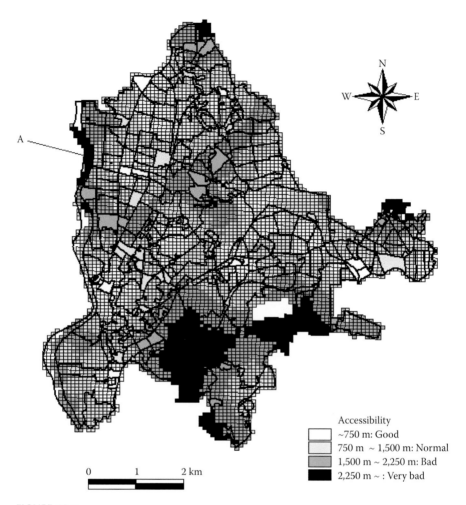

FIGURE 19.7
Composite map of accessibility at residential district level and 100 m mesh level.

100m meshes consisting of it, the district is considered as a uniform area in terms of accessibility. In other words, no variability exists within the residential district. Then the analytical results at the district level are used. Contrarily, if a district includes 100 m meshes with different accessibility levels, the results at 100 m mesh level will be used, because spatial variation of the accessibility cannot be represented at the district level.

Figure 19.7 shows the accessibility composed at two levels, depending on its spatial variability. In 261 districts[2], the residential districts with homogeneous accessibility are 49 (18.8 percent). Namely, accessibility measured at the district level has enough accuracy for about 20 percent of districts. The accessibility levels for their districts are broken down into 12 districts (24.5 percent) as "good," 20 (40.8 percent) as "normal," 15 (30.6

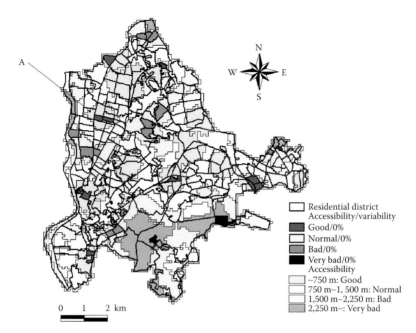

FIGURE 19.8
(See color insert following page 176.) Accessibility map at variable spatial level.

percent) as "bad," and 2 (4.1 percent) as "very bad." In comparing the composition ratio of each level for all districts[3], appearance of the districts with homogeneous accessibility is somewhat low to "normal" level and slightly high to "bad" level.

This visualization can show where and to what degree accessibility differs inside the district. However, this may not be considered as an efficient visualization technique, because 100 m mesh level is too high a spatial resolution in order to represent spatial variation of accessibility within the district

19.2.3.3 Accessibility Map at Variable Spatial Level

The third visualization represents an analytical result at the suitable level for its accuracy. While the districts with homogeneous accessibility are represented at the district level, contiguous 100 m meshes with the same accessibility level are aggregated to a homogeneous area. This is done by dissolve processing on GIS. Figure 19.8 shows the distribution of accessibility by the visualization at the variable spatial level. This efficiently represents the distribution of accessibility between and within the districts. The districts with "good" access appear as pink, the districts with "normal" as yellow, the districts with "bad" as sky-blue, and the districts with "very bad" as blue. The districts with homogeneous access are highlighted in each color and are bounded by a thick line. In addition, uniform areas are formed according to spatial variation of accessibility at 100 m mesh level.

Because the results obtained from spatial analysis (measure of accessibility) have to be used with some reliability in the practical scenes, GIS should perform multilevel analysis and adopt various visualization techniques to represent the analytical result according to its accuracy.

19.3 Measure of Accessibility by Highly Accurate Simulation and Its Visualization

This section considers site assessment in terms of accessibility for retail and service facilities. The case study here is based on a site-assessment problem proposed by a car dealer. This dealer opened a new outlet at site A along Circle Road 8 in Setagaya Ward, Tokyo, which is shown as a symbol "A" in Figure 19.9, but its sales results was poorer than the one the dealer expected. The competitor holds site B along the same road, as shown in Figure 19.9. Its sales results are said to be fairly good. The dealer wants to know the reason why site A is not good for sales, though it is facing the same road in the same ward as site B.

19.3.1 Population as Demand Volume

To measure demand volume at the first step of site assessment (Geertman et al., 2004), the buffer/overlay approach was applied to sites A and B to calculate the populations within circle zones of 1- and 2-kilometer radiuses by direct distance from sites A and B (Figure 19.9). The populations at site A are about 32,000 in the 1 km zone and 140,000 in the 2 km zone (Table 19.2a), and the populations at site B are 56,000 and 160,000, respectively (Table 19.2b). From this result, it is said that the population in each zone at site A is nearly 20,000 smaller than that at site B.

The population densities for the 2 km zone at sites A and B are about 11,000/km^2 and 12,000/km^2, respectively, and then are slightly lower than average population density, 13,000/km^2, for 23 wards of Tokyo. The reason why site A has a slightly smaller demand (population) than site B or the average area in Tokyo wards is thought that the zone at site A includes Tama River (Figure 19.9). Therefore the dull business at site A may be explained, in some degree, by this fact of smaller demand volume by 20,000.

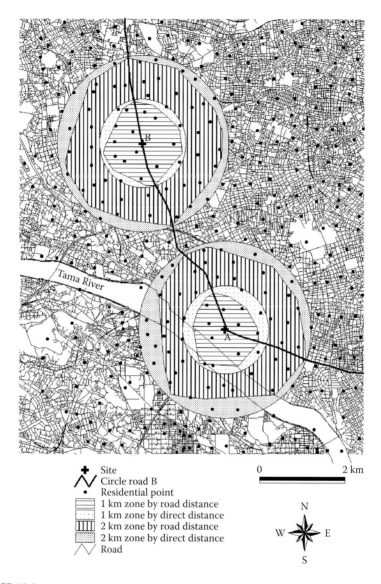

FIGURE 19.9
Direct-distance zones and road-distance zone of 1 km and 2 km around sites A and B.

19.3.2 Development of Road Network

Next, the accessibility to the populations in 1 km and 2 km zones from the site is analyzed in terms of the development of road network. Figure 19.9 also shows the zones of 1 km and 2 km at sites A and B, defined by road distance using the "Find service area" tool in the network analysis of ArcView. In comparison with the areas of 1 km and 2 km zones defined

TABLE 19.2a

Direct distance		
	1km zone	2km zone
Area	3.1km2	12.6km2
No. of district	15	59
Population	32,171	140,009
Population density	10,378/km2	11,112/km2

Road distance		
	1km zone(%)	2km zone(%)
Area	1.9km2(61.3%)	9.5km2(75.6%)
No. of district	9(60.0%)	46(78.0%)
Population	21,628(67.2%)	102,428(73.2%)

Navigation road distance		
	1km zone(%)	2km zone(%)
Area	0.7km2(22.6%)	2.8km2(22.2%)
No. of district	4(26.7%)	9(15.3%)
Population	7,302(22.7%)	23,298(16.6%)

TABLE 19.2b

Direct distance		
	1km zone	2km zone
Area	3.1km2	12.6km2
No. of district	18	62
Population	56,139	160,546
Population density	18,109/km2	12,742/km2

Road distance		
	1km zone(%)	2km zone(%)
Area	2.4km2(77.4%)	10.2km2(81.0%)
No. of district	13(72.2%)	51(82.3%)
Population	39,159(69.8%)	135,198(84.2%)

Navigation road distance		
	1km zone(%)	2km zone(%)
Area	1.4km2(45.2%)	6.6km2(52.4%)
No. of district	4(22.2%)	26(41.9%)
Population	19,272(34.3%)	95,512(59.5%)

by direct distance, the areas of 1 km and 2 km zones by road distance occupy 61.3 percent and 75.6 percent, respectively, at site A (Table 19.2a) and 77.4 percent and 81.0 percent respectively at site B (Table 19.2b). Figure 19.9 shows the centroids (●) of residential districts (Cyocyo) within 1 km and 2 km zones defined by direct and road distances. Numbers of residential points within 1 km and 2 km zones defined by road distance are 60 percent and 78 percent, respectively, at site A and 72.2 percent and 82.3 percent, respectively, at site B, in comparison with the numbers within 1 km and 2 km zones by direct distance.

Thus, the coverage ratios of 1 km and 2 km zones by road distance to ones by direct distance are about 60 percent and 80 percent, respectively, at site A and about 70 percent and 80 percent, respectively, at site B. From this fact, site A is inferior to site B in the development of a road network.

As a result, the population in the 1 km zone based on road distance at site A is 21,000, and that in the 2 km zone is 102,000 (Table 19.2a). The same populations at site B are 39,000 in the 1 km zone and 135,000 in the 2 km zone (Table 19.2b). Therefore, the dull business at site A may be explained by a smaller demand than site B by 33,000 in the 2 km zone, in addition to taking into consideration less development of the road network.

19.3.3 Measure of Navigation Road Distance by Highly Accurate Simulation Considering Complex Traffic Conditions

Analyzing accessibility by car, narrow road, one-way road, traffic signals, median strips, and toll road makes traffic conditions complex in the metropolitan area. Network analysis on such a complex road network cannot search actual shortest path. Therefore, a highly accurate simulation under complex traffic conditions is performed to find a navigation route as an actual

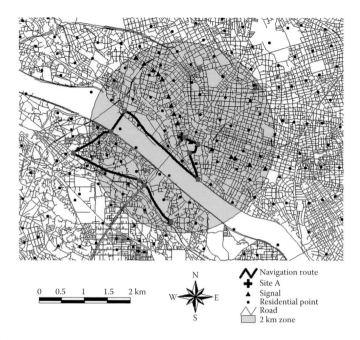

FIGURE 19.10a
Navigation route to site A in considering Tama River.

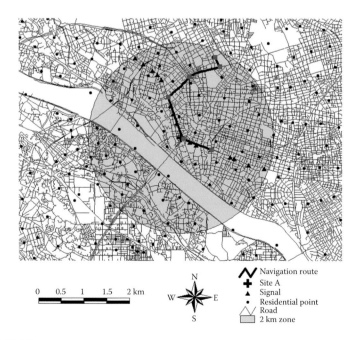

FIGURE 19.10b
Navigation route to site A in considering U-turn.

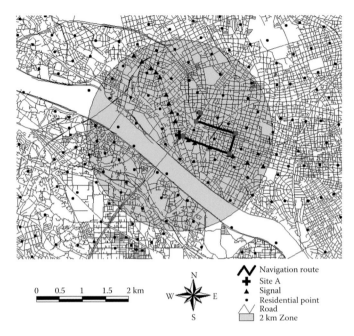

FIGURE 19.10c
Navigation route to site A in considering narrow road and one-way road.

shortest path by interacting between man and machine, namely, the "Shortest path finder" tool in ArcView.

Three obstacle factors are considered in a highly accurate simulation of navigation routes to prove the inaccessible condition at site A. The first obstacle factor is Tama River, consisting of the boundary between the metropolis of Tokyo and Kanagawa Prefecture. Figure 19.10a shows a navigation route for the resident in Kanagawa Prefecture. They will cross Tama River by using "Futako" bridge instead of using "Daisan-keihin" as a toll road. Therefore, they make a detour, as shown in Figure 19.10a.

The second obstacle factor is that the site faces on the outer (northward) side of Circle Road 8, which is one of industrial arterial roads in Tokyo. The separation of the outer and inner sides by median strips makes site A inaccessible in turning right directly on Circle Road 8. Then the customer approaching from the inner side has to pass the site and make a U-turn at the intersection 700 meters away, as shown in Figure 19.10b.

The third factor is that there are many narrow roads and one-way roads in Setagaya Ward. The road network, with many narrow and one-way roads, obstructs the passing of cars. Figure 19.10c shows an example of a detour for the neighboring resident to approach site A.

The navigation road distance from each residential point to sites A and B were measured by a highly accurate simulation considering complex traffic conditions. Figure 19.11 shows the accumulated population as a vertical axis

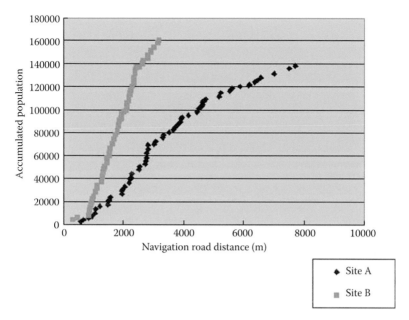

FIGURE 19.11
Navigation road distance as horizontal axis versus accumulated population as vertical axis.

versus the navigation road distance to sites A and B shown as a horizontal axis. While the residential points located in the 2 km zone at site B are limited within 3 kilometers in terms of navigation road distance, the residential points in the 2 km zone at site A are extended to 8 kilometers in terms of navigation road distance.

The isoline map shown in Figure 19.12a visualizes the distribution of navigation road distance for each residential point in a 2 km zone by direct distance from site A. In considering complex traffic conditions, the 1 km zone by navigation road distance becomes small size, and the 2 km zone is extended in a southeast direction. The areas of 1 km and 2 km zones by navigation road distance only occupy 22 percentage of 1 km and 2 km zones by direct distance (Table 19.2a).

Figure 19.12b shows the distribution of navigation road distance for each residential point in the 2 km zone by direct distance from site B. In comparison with site A, 1 km and 2 km zones by navigation road distance are expanded to all direction. One-kilometer and 2-kilometer zones by navigation road distance occupy 40 percent to 50 percent of these zones by direct distance (Table 19.2b).

The population within 1 km and 2 km zones by navigation road distance are 7000 (22 percent of the population in the 1 km zone by direct distance) and 23,000 (16 percent) at site A, and 19,000 (34 percentage) and 95,000 (about 60 percent) at site B, as shown in Tables 19.2a and 19.2b.

As the result of analysis by highly accurate simulation, it becomes clear that site A has 73,000 smaller demand in the 2 km zone than site B. The dull

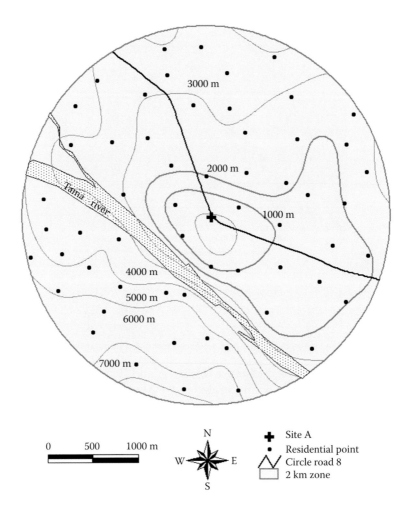

FIGURE 19.12a
Isoline map of navigation road distance from site A.

business at site A can be explained at last by performing a highly accurate simulation.

19.4 Conclusion

GIS is expected to be a useful tool for site assessment. However, as GIS treats geographic features at the district (community) level or larger spatial levels, the analysis at point (site) level may step forward to an unknown stage. To apply site assessment to the practical scene, it is necessary to improve the accuracy of spatial analysis in terms of the following three aspects.

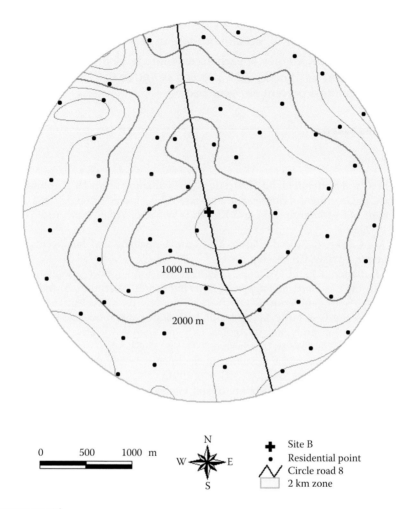

1000 m

2000 m

N

W �֍ E

S

0   500   1000 m

✚ Site B
• Residential point
⋀⋁ Circle road 8
▢ 2 km zone

FIGURE 19.12b
Isoline map of navigation road distance from site B.

The first is to reproduce a more detailed environment of GIS. More realistic solutions could be derived by reproducing complex traffic conditions, as shown in the third section of this paper. The second is to perform spatial analysis at a higher resolution level as 100 m mesh level shown in the second section. The third is related to building a more powerful spatial model that incorporates many variables. The research from the above three aspects should be furthered so that the result of spatial analysis holds accuracy levels enough for practical use.

1. The Basic Area Mesh System covers the whole of Japan with 1 km meshes.

2. The districts here consist of residential districts and their detached districts, which are 10 hectare or more in area.

3. The composition ratios of each accessibility level for all districts are 24.8 percent as "good," 47.2 percent as "normal," 23.6 percent as "bad," and 4.4 percent as "very bad."

References

Beard, M.K. and Buttenfield, B.P., Detecting and evaluating errors by graphical methods, in *Geographical Information Systems*, Vol. 1, 2nd ed., Longley, P.A., Goodchild, M.F., Maguire, D.J., and Rhind, D.W., Eds., Wiley, New York, 1999, p. 219–233.

Birkin, M., Clarke, G.P., and Clarke, M., *Retail Geography and Intelligent Network Planning*, Wiley, Chichester, 2002.

Birkin, M., Boden, P., and Williams, J., Spatial decision support systems for petrol forecourts, in *Planning Support Systems in Practice*, Geertman, S. and Stillwell, J., Eds., Springer, Berlin, 2003.

Birkin, M., Clarke, G., Clarke, M., and Culf, R., Using spatial models to solve difficult retail location problems, in *Applied GIS and Spatial Analysis*, Stillwell, J. and Clarke, G., Eds., Wiley, Chichester, 2004, pp. 35–54.

Fisher, P., Visualization of the reliability in classified remotely sensed images, *Photogramm. Eng. Remote Sens.*, 60, 905–910, 1994a.

Fisher, P., Visualizing the uncertainty of soil maps by animation, *Cartographica*, 30, 20–27, 1994b.

Gahegan, M., Visualization as a tool for GeoComputation, in *GeoComputation*, Openshaw, S. and Abrahart, R.J., Eds., Taylor & Francis, London, 2000, pp. 253–274.

Geertman, S., Jong, T. de, Wessels, C., and Bleeker, J., The relocation of ambulance facilities in central Rotterdam, in *Applied GIS and Spatial Analysis*, Stillwell, J. and Clarke, G., Eds., Wiley, Chichester, 2004, pp. 215–232.

Jones, K. and Simmons, J., *The Retail Environment*, Routledge, London, 1990.

Kwan, M., Space-time and integral measures of individual accessibility: a comparative analysis using a point-based framework, *Geogr. Anal.*, 30(3), 191–216, 1998.

MacEachren, A.M., *How Maps Work: Representation, Visualization, and Design*, Guilford Press, New York, 1995.

Orford, S., *Valuing the Built Environment: GIS and House Price Analysis*, Ashgate, Hants, U.K., 1999.

Sekine, T., Analysis of spatio-temporal stability in accessibility: a case study ophthalmic hospitals in Matsudo City, Chiba Prefecture, *Geogr. Rev. Jpn.*, 76(10), 725–742, 2003.

Talen, E. and Anselin, L., Assessing spatial equity: an evaluation of measures of accessibility to public playgrounds, *Environ. Plann. A*, 30, 595–613, 1998.

Van der Wel, F.J.M., Hootsmans, R.M., and Ormeling, F., Visualization of data quality, in *Visualization in Modern Cartography*, MacEachren, A.M. and Taylor, D R.F., Eds., Elsevier, Oxford, U.K., 1994, pp. 313–331.

20

Visualization of the Mental Image of a City Using GIS

Yukio Sadahiro and Yoshio Igarashi

CONTENTS

20.1 Introduction

Visualization is one of the essential functions of Geographical Information System (GIS) (Cromley, 1992; MacEachren and Taylor, 1994; Nielson et al., 1997; Slocum, 1998). As a tool of spatial analysis, it is an efficient way to explore spatial phenomena. We often grasp the structure of a spatial phenomenon by only looking at the picture indicating the phenomenon. Changing the scale of visualization, we detect spatial patterns at various scales from local to global. Visualization is also useful for making a decision on spatial phenomena. In sightseeing, for instance, tourist maps help us finding good places to visit and stay. Bus-route maps tell us which routes we need in order to reach our destinations. Crime maps show us the regional variation of crime rate — how dangerous it is to visit a certain place. Weather maps are indispensable in making plans for a field trip.

As well as physical and concrete objects, abstract information can also be visualized in GIS if represented as a computational model. To explore a wider application of GIS, this paper discusses the visualization of an abstract concept, the mental image of a city, with a focus on its spatial variation. The image of a city is usually communicated by text information, typically a sentence characterizing a location by adjectives. We may say, "That square is lively and often bustling," "The art galleries and antique shops create an artistic atmosphere on the street," and "The downtown area is very calm, so I sometimes feel it is dangerous." The objective of this paper is to incorporate these literal representations into GIS to visualize the image of a city.

In academics the mental image of a city is often discussed in architecture and environmental psychology (Bell et al., 1990; Bechtel and Churchman, 2002). Psychologists are interested in the relationship between the image of a space and its physical elements, such as buildings, roadways, and pavements, to understand the structure and formation of mental image. Architects look at this relationship from a more practical viewpoint, that is, how to give a good impression to visitors of a space. Visualization of the image of a city would help in studying the relationship between physical and mental spaces.

Image visualization is also useful in marketing and traveling. Image is critical in apparel industries. When locating a new store, a company examines the image of a city in detail to seek the best location for not only selling its products, but also improving the image of the company and its brands. When we visit a new city, we often wish to stroll around the city rather than visit certain places. In such a case, it is useful to know the image of streets and regions of a city rather than detailed information of individual facilities. Individual regions in New York, say, SOHO, East Village, and Harlem, are characterized by their own images, which helps visitors of New York understand the urban structure of New York and make a trip plan.

As mentioned above, the image of a city is usually represented as text information, which cannot be directly treated in GIS. To incorporate such information into GIS, we first describe the formal representation of the image in the following section. We then discuss how the image is created by spatial objects, which leads to a mathematical model of the image. The section ends with discussion on the visualization methods of the image in GIS. Section 20.3 shows a prototype system that visualizes the image of a city, taking Shibuya in Tokyo, Japan, as an example. Source data, a model of the image, and a visualization method are described in turn, which is followed by the system evaluation by users. Section 20.4 summarizes the features of the system with discussion for further research.

20.2 Methodology

20.2.1 Representation of the Image of a City

The image of a city is usually described by adjectives, say, lively, bustling, busy, sophisticated, calm, lonely, and dangerous, often with adverbs, such as extremely, considerably, very, moderately, and slightly. This implies that the image consists of numerous elements represented by adjectives. We thus define the image of a city as a set of elements, each of which is a function of location, time, and individual. Take, for instance, the liveliness of a city. Since the liveliness varies from place to place and changes over time, it is reasonable to assume a function of location and time. It also varies among individuals because it happens that some feel lively while others do not in the same situation.

The above definition is described mathematically as follows. Assume that the image of a city of region S consists of m elements, such as the liveliness, calmness, and dangerousness. Given a location $\mathbf{x}$ and a time t, we denote the perceptual degree of element i by an individual j as $f_{ij}(\mathbf{x}, t)$. The image of a city is then represented as a set of functions $F = \{f_{ij}(\mathbf{x}, t), i = 1, \ldots, m, j = 1, \ldots, n\}$.

This representation allows variations in three dimensions, that is, location, time, and individual. This high flexibility, though it seems quite reasonable, makes it difficult to visualize the image of a city as it is in GIS. Even if we fix the time at t, we still have $m'n$ distributions to visualize. It is difficult to understand the structure of the image if we visualize them in GIS as they convey too much information about the image. To reduce the amount of information, we summarize the variation among individuals by their mean and variance. We replace $F_i = \{f_{ij}(\mathbf{x}, t), j = 1, \ldots, n\}$, the set of functions of element i, by their mean $m_i(\mathbf{x}, t)$ and variance $s^2_i(\mathbf{x}, t)$. The image of a city is then represented by a set of functions $I = \{m_i(\mathbf{x}, t), s^2_i(\mathbf{x}, t), i = 1, \ldots, m\}$.

20.2.2 Model Description

Having defined the representation of the image of a city, we then propose its mathematical model. The image of a city at a certain location depends on the properties of its surrounding spatial objects. For instance, the image of a square is determined by buildings, streets, sidewalk stands, and so forth. The effect of a spatial object usually decreases with the distance from its location. A beautiful building greatly improves the image of its surrounding area, while it rarely affects the image of a distant place. These observations naturally give a mathematical model of the image defined as follows.

Suppose K spatial objects with L properties distributed in S. The location of spatial object k is denoted by $\mathbf{z}_k$. The property l of spatial object k at time t is $a_{kl}(t)$. The mean of image element i at $(\mathbf{x}, t)$ is given by

$$\mu_i(\mathbf{x},t)=\frac{1}{K}\sum_k\sum_l \rho_{il}\left(\left|\mathbf{x}-\mathbf{z}_k\right|\right)g_i\left(a_{kl}(t)\right) \qquad (20.1)$$

where $g_i(a_{kl}(t))$ is the effect of property l of spatial object k on element i, and $r_{il}(|\mathbf{x}-\mathbf{z}_k|)$ is its distance-decay function.

The variance of the image among individuals also depends on the properties of surrounding spatial objects. This paper assumes that it is a function of the variance in the effect of spatial objects and that it decreases with the number of spatial objects:

$$\sigma_i^2(\mathbf{x},t)=\frac{1}{\displaystyle\sum_k v\left(\left|\mathbf{x}-\mathbf{z}_k\right|\right)}\cdot h\left(\frac{1}{K}\sum_k\left\{\sum_l \rho_{il}\left(\left|\mathbf{x}-\mathbf{z}_k\right|\right)g_i\left(a_{kl}(t)\right)-\mu_i(\mathbf{x},t)\right\}^2\right)$$

$$(20.2)$$

where $n(|\mathbf{x}-\mathbf{z}_k|)$ is a distance-decay function. The latter assumption implies that the image is consistent among individuals where many spatial objects are clustered; individuals receive more information with an increase of spatial objects, which makes the image clearer.

Specifying the functions $r_{il}(|\mathbf{x}-\mathbf{z}_k|)$, $g_i(a_{kl}(t))$, $n(|\mathbf{x}-\mathbf{z}_k|)$, and $h(\mathbf{x},t)$, we obtain a mathematical model of the image of a city with some unknown parameters. These parameters are usually estimated through a questionnaire survey. A typical method is to ask subjects to rate each element of the image at sample locations and fit the model to the result obtained. An example of model estimation will be shown later.

20.2.3 Visualization of the Image of a City

Once a model is estimated, the image of a city is visualized in GIS. A direct and straightforward method is to build computational models of the function set I in GIS, such as Triangular Irregular Networks (TINs) and lattices, and visualize them as three-dimensional surfaces. Along with this ordinary method, this paper proposes smoothing of the functions. When interests lie only in the outline of the image, details are not necessary or even redundant, because they conceal the global structure of the image and their visualization takes considerable time even if a high-performance computer is employed.

The smoothing operation used in visualization is spatially inhomogeneous, that is, it depends on the density of spatial objects. The smoothing function keeps the details of functions where spatial objects are densely distributed, while it makes them smooth where spatial objects are sparse. This is because we are interested in the local variation of the image where spatial objects are clustered. The smoothing operation on $f(\mathbf{x})$ is mathematically defined by

$$s(\mathbf{x}) = \int_{\mathbf{y} \in S} \exp\left[-\left\{ \gamma + \kappa \sum_k \mathrm{v}\left(|\mathbf{x} - \mathbf{z}_k|\right) \right\} |\mathbf{x} - \mathbf{y}| \right] f(\mathbf{y}) d\mathbf{y} \qquad (20.3)$$

Parameters g and k determine the scale of smoothing. The former g is an ordinary smoothing parameter; a large g yields smooth surfaces. The latter k, on the other hand, gives the spatial variation of smoothing by using the term

$$\sum_k \mathrm{v}\left(|\mathbf{x} - \mathbf{z}_k|\right),$$

the density of spatial objects around location $\mathbf{x}$. A large k gives more details where spatial objects are clustered; if k is zero, smoothing operation is homogeneous in S.

Consequently, the mean and variance of image element i at $(\mathbf{x}, t)$ are visualized as surfaces defined by

$$\mu_i{}'(\mathbf{x}, t) = \int_{\mathbf{y} \in S} \exp\left[-\left\{ \gamma + \kappa \sum_k \mathrm{v}\left(|\mathbf{x} - \mathbf{z}_k|\right) \right\} |\mathbf{x} - \mathbf{y}| \right] \mu_i(\mathbf{y}, t) d\mathbf{y}$$

$$= \int_{\mathbf{y} \in S} \exp\left[-\left\{ \gamma + \kappa \sum_k \mathrm{v}\left(|\mathbf{x} - \mathbf{z}_k|\right) \right\} |\mathbf{x} - \mathbf{y}| \right] \sum_k \sum_l \rho_{il}\left(|\mathbf{y} - \mathbf{z}_k|\right) g_i\left(a_{kl}(t)\right) d\mathbf{y}$$

$$= \sum_k \sum_l g_i\left(a_{kl}(t)\right) \int_{\mathbf{y} \in S} \exp\left[-\left\{ \gamma + \kappa \sum_k \mathrm{v}\left(|\mathbf{x} - \mathbf{z}_k|\right) \right\} |\mathbf{x} - \mathbf{y}| \right] \rho_{il}\left(|\mathbf{y} - \mathbf{z}_k|\right) d\mathbf{y}$$

$$(20.4)$$

and

$$\sigma_i^{2\,\prime}(\mathbf{x}, t) =$$

$$\int_{\mathbf{y} \in S} \exp\left[-\left\{\gamma + \kappa \sum_k v\left(\left|\mathbf{x} - \mathbf{z}_k\right|\right)\right\} \left|\mathbf{x} - \mathbf{y}\right|\right] \sigma_i^2(\mathbf{y}, t) d\mathbf{y}$$

$$= \int_{\mathbf{y} \in S} \frac{\exp\left[-\left\{\gamma + \kappa \sum_k v\left(\left|\mathbf{x} - \mathbf{z}_k\right|\right)\right\} \left|\mathbf{x} - \mathbf{y}\right|\right] h\left(\frac{1}{K} \sum_k \left\{\sum_l \rho_{il}\left(\left|\mathbf{x} - \mathbf{z}_k\right|\right) g_i\left(a_{kl}(t)\right) - \mu_i(\mathbf{x}, t)\right\}^2\right)}{\sum_k v\left(\left|\mathbf{y} - \mathbf{z}_k\right|\right)} d\mathbf{y}$$

(20.5)

respectively.

Given an element i and a time t, the image of a city is represented by a pair of two-dimensional distributions defined by the above two equations. They are usually visualized as two surfaces in GIS. In theory, however, we can visualize four distributions simultaneously by a single surface, because we have three elements of color — hue, saturation, and brightness — as well as surface height, to indicate function values. For instance, we may show the mean and variance of a certain element simultaneously by using the height and brightness of a single surface. The mean of two elements can be visualized by the height and saturation of a surface. Though care should be taken in the choice of visualization method, it is evident that functions of GIS extend the potential for visualizing spatial phenomena.

20.3. A Prototype System

To implement the method proposed, we built a prototype system using GIS. The study area is Shibuya in Tokyo, Japan, a major subcenter of Tokyo primarily composed of business districts and commercial areas. Shibuya station is one of the biggest railway stations in Tokyo, which has 2 million passengers per day. The objective of the system is to visualize the spatiotemporal distribution of the image of Shibuya area.

20.3.1 Spatial Data

To describe the image of Shibuya, we used spatial data of restaurants, because Shibuya is characterized by large commercial areas that attract a wide variety of people, from young to aged. We obtained a list of restaurants from a Web site, *Gourmet Pia* (Pia, 2003). The Web site provides the list of restaurants with their attributes, such as the location, cuisine type, price range, and hours, as well as the attributes of customers, including age distribution, group size, and male/female ratio. The Web site also rates the atmosphere of restaurants on several dimensions, such as cheerfulness and calmness on a scale from one to five. We converted the addresses into spatial data by geocoding, and linked their attributes to the spatial data.

20.3.2 Model of the Image of Shibuya

Following the method proposed in the previous section, we represent the image of Shibuya by a set of elements. To choose important elements, we applied principal-component analysis (Johnson and Wichern, 2002; Anderson, 2003) to the restaurant evaluation rated by the Web site. The analysis yielded two principal components, which we call *liveliness* and *elegance*, represented as two pairs of functions $\{m_1(\mathbf{x}, t), s_1^2(\mathbf{x}, t)\}$ and $\{m_2(\mathbf{x}, t), s_2^2(\mathbf{x}, t)\}$, respectively.

The definition of these functions is given by Equations 20.1 and 20.2. As seen in the equations, the definition requires specification of the functions $r_{il}(|\mathbf{x} - \mathbf{z}_k|)$, $g_i(a_{kl}(t))$, $n(|\mathbf{x} - \mathbf{z}_k|)$, and $h(\mathbf{x}, t)$. The function $g_i(a_{kl}(t))$ is naturally derived from the principal-component analysis. As for the function $h(\mathbf{x}, t)$, we assume that it depends on neither location $\mathbf{x}$ nor the time t for simplicity. The distance-decay functions are defined as

$$\rho_{il}\left(\left|\mathbf{x} - \mathbf{z}_k\right|\right) = \exp\left(-\left|\mathbf{x} - \mathbf{z}_k\right|\right) \tag{20.6}$$

and

$$v\left(\left|\mathbf{x} - \mathbf{z}_k\right|\right) = \exp\left(-\alpha\left|\mathbf{x} - \mathbf{z}_k\right|\right) \tag{20.7}$$

where a is an unknown parameter to be estimated. Equations 20.1 and 20.2 then become

$$\mu_i\left(\mathbf{x}, t\right) = \sum_k \exp\left(-\left|\mathbf{x} - \mathbf{z}_k\right|\right) \sum_l g_i\left(a_{kl}\left(t\right)\right) \tag{20.8}$$

and

$$\sigma_i^2\left(\mathbf{x}, t\right) = \frac{1}{\sum_k \exp\left(-\alpha\left|\mathbf{x} - \mathbf{z}_k\right|\right)} \tag{20.9}$$

respectively.

Unlike Equation 20.8, Equation 20.9 contains an unknown parameter a. To estimate it, we conducted an experiment in the Department of Urban Engineering at the University of Tokyo. Twenty-five graduate students served as subjects who were naive as to the purpose of the experiment. In the experiment, we showed a map of Shibuya to the subjects, on which circles the radius of 200 meters were drawn. We asked them to evaluate the clearness of the image of each circular region on a scale from one (very ambiguous) to five (very clear). From the observed data we estimated the model given by Equation 20.9 using the least-square method to obtain a = –0.0137, which is statistically significant at the 5 percent level.

20.3.3 Visualization of the Image of Shibuya

Having obtained the model of the image, we visualized it using ArcGIS 8.1 with a visualization package AVS/Express 6.0 (for details, see Igarashi 2003). The system visualizes the two elements of the image of Shibuya, liveliness and elegance, as continuous surfaces. The mean of an image element is indicated by both the height and hue of a surface, while the variance is indicated by the brightness. Users determine the details of visualization method through a graphic interface (Figure 20.1): location, direction, scale, time, and surface color, as well as smoothing parameters g and k. Figure 20.2 shows examples of the image of Shibuya visualized by the system.

The system utilizes the inhomogeneous smoothing in visualization. As seen in Figure 20.3, the image is shown in detail around Shibuya station where restaurants are clustered so that users can see the local variation of the image. On the other hand, users can grasp the global structure of the image where restaurants are dispersed.

The Web site Gourmet Pia shows the opening hours of restaurants in Shibuya, as mentioned earlier. The data permit the system to visualize the change of the image over time. Assuming that closed restaurants do not affect the image, the system fixes the function value $g_i(a_{kl}(t))$ at zero, while restaurant k is closed and calculates the image elements. Figure 20.4 shows the elegance of Shibuya in the daytime and nighttime, which shows a distinct difference.

Calculation of the image may take time on a classic computer, and, consequently, visualization of its change on demand may be irritating. However, the system can store the results of calculations as a single movie file; we can see the change of the image as a movie at a reasonable speed even in an insufficient computer environment.

20.3.4 System Evaluation

To seek evaluations by users on the system, we conducted a questionnaire survey. Twelve graduate students in the Department of Urban Engineering at the University of Tokyo, who were familiar with the Shibuya area, explored the image of Shibuya using the system. They learned the operation of the system by the hard-copy manual. We asked them to evaluate the

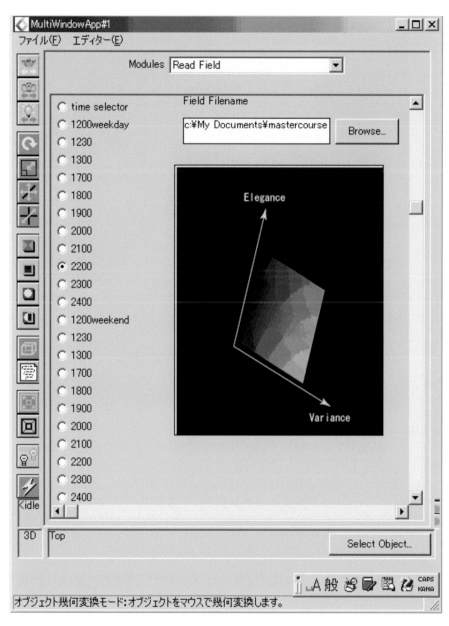

FIGURE 20.1
User interface of the system.

system in terms of 1) operability of user interface and 2) agreement between the image that they have in mind and that visualized by the system.

The user interface received favorable opinions from most of the respondents. They stated that they could learn the operation of the system only within a few minutes. Adoption of slide bars was highly evaluated.

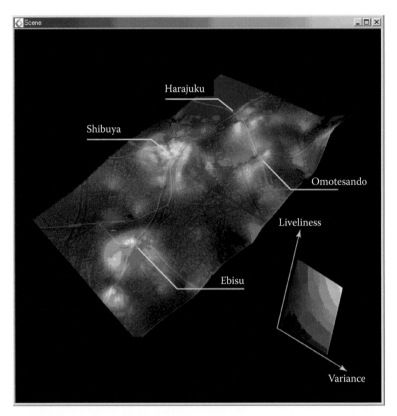

FIGURE 20.2a
(See color insert following page 176.) The image of Shibuya: a) liveliness and b) elegance.

Evaluation of the image visualized by the system varies among locations. In general, the image visualized was close to that of the respondents where restaurants are clustered. This supports the assumption about the variance of the image among individuals mentioned in the previous section: Clustering of spatial objects makes the image more consistent among individuals.

Besides the answers to our questions, respondents gave us some additional comments on the system. The main purpose of the system is to communicate the image of a city to those who are not familiar with the city, and many respondents stated that the system has achieved this goal. In addition, some suggested another use for the system. They stated that the image visualized reminds them of the details of the city, say, the atmosphere of each restaurant or street. This implies that the system is useful also for those familiar with the city when making a trip plan or choosing a restaurant, because the system extends their choice options.

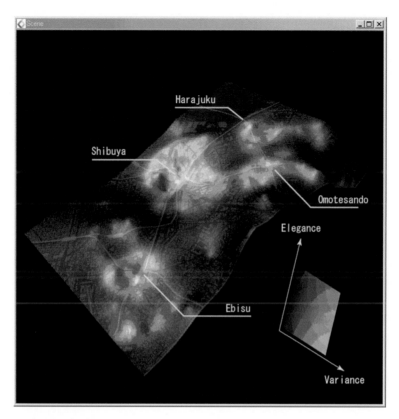

FIGURE 20.2b (continued)
(See color insert following page 176.)

20.4 Conclusion

In this paper we have proposed a method for visualizing an abstract concept, the mental image of a city, with a focus on its spatial variation. We represented the image of a city as a mathematical model that can be calculated from spatial data widely available on the Internet. Using the method, we built a GIS-based system that visualizes the image of Shibuya in Tokyo, Japan.

Advantages of the system are summarized as follows.

The System Visualizes the Spatial Distribution of an Abstract Concept, the Image of a City. This paper shows a method for visualizing the image of a city. It is a good example of treating an abstract rather than a concrete spatial concept in GIS, and, consequently, suggests potential wider applications of GIS to human and social sciences where abstract spaces are more frequently discussed. Social, mental, and cultural spaces may be naturally handled within GIS along with physical space in the future.

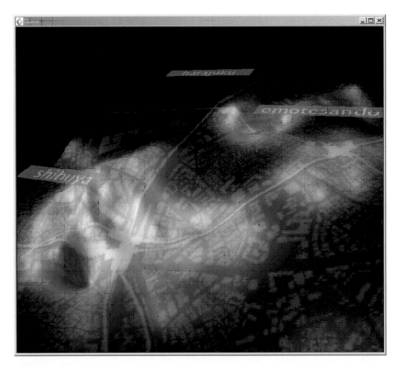

FIGURE 20.3
(See color insert following page 176.) The image of Shibuya.

The Image is Presented as a Surface, a Continuous Spatial Distribution. Spatial objects in a city are usually represented in discrete forms, that is points, lines, and polygons. Consequently, the spatial distributions of their attributes, including the image, are also visualized in discrete forms. Discrete representation, however, is not appropriate for visualizing the image of a city, because the image is ambiguous and subjective to some extent so that it is spatially smooth without sudden changes. Conversion of discrete spatial objects into a surface allows more realistic visualization of the image.

The System Utilizes Inhomogeneous Smoothing in Image Visualization. This is a unique function of the system, which enables us to see the outline and details of the image simultaneously. The degree of details presented depends on the density of spatial objects; the system shows the details where spatial objects are clustered, while it makes the image surface smoother where objects are sparse. Users do not have to change the scale of visualization when looking at a different place of a different object density.

Visualization is Performed in the Spatiotemporal Domain. The system visualizes the change of the image over time, as well as its spatial distribution. This is critical, because the image of a city often changes drastically between day and night. Spatiotemporal data necessary for visualizing the change were not widely available, especially in a digital format. Fortunately, with the spread of spatial data, digital data of temporal information has also

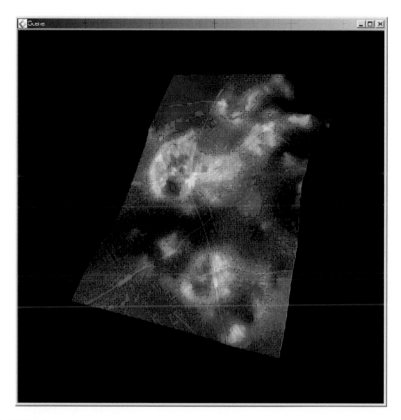

FIGURE 20.4a
(See color insert following page 176.) Elegance of Shibuya in the a) daytime and b) nighttime.

become available, sometimes on the Internet, as shown in this paper. This greatly reduces the cost of system construction and, consequently, extends the applicability of the system.

The Spatial Data are Generated from the Information Available on the Internet. The Internet is rapidly growing as an inexhaustible source of spatial information. It provides various information about spatial objects other than restaurants, such as retail stores, theaters, museums, and streets. We can easily improve the image visualized by taking these spatial objects into account in model building. The close connection to the Internet also implies an automatic update of spatial data and that of the image of a city calculated from the data. This is a great advantage for computer-based systems, such as GIS, that treat massive and latest spatial data, because manual update of spatial data is very costly, which severely limits the applicability of the system.

For further extention of the system, we still have many problems to resolve. For instance, further discussion is necessary on the choice of spatial data and spatial model for GIS applications in human and social sciences. Spatial data and model are both highly dependent on the field to which GIS is

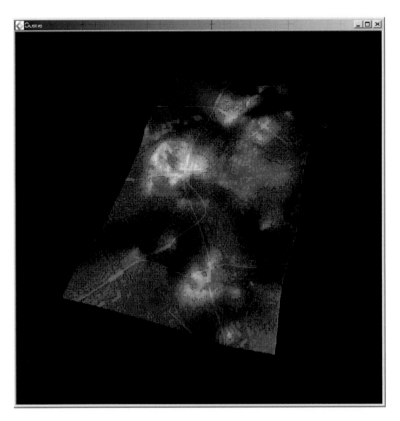

FIGURE 20.4b (continued)
(See color insert following page 176.)

applied. Therefore, it is critical to choose spatial data and model appropriate for a specific space that needs spatial analysis and visualization. An extensive and general discussion on this topic is indispensable for a wide spread of GIS in human and social sciences. Automatic update of spatial data and the image calculated from the data requires a system that extracts necessary information from the Internet. It is somewhat simple if the data source is one specific site that provides spatial data in a systematic form. However, if information stored in nonsystematic ways has to be gathered from multiple sites, it is necessary not only to develop different interfaces for individual sites but also to integrate the information that may be inconsistent with each other. This is a challenging topic in the GIS community.

Along with these extensions, theoretical basis of the system should be discussed further. In the prototype system, we adopted a rather simple model to represent the image of Shibuya. This is because estimation of simple models requires only a small amount of spatial data so that the cost of data collection and model estimation is not expensive. On the other hand, simple models are less realistic than more complicated models that take into account various factors affecting the image. An efficient method of data acquisition and model estimation should be considered in future research. The visualization system also has a room for improvement. The present system totally leaves the choice of visualization method, say, scale and colors, to users. Though it allows high flexibility, some users may feel irritated to spend time visualizing the image as they wish. A more intelligent system that chooses an appropriate method of visualization automatically should be explored.

Literature Cited

Anderson, T.W., *An Introduction to Multivariate Statistical Analysis*, John Wiley & Sons, New York, 2003.

Bell, P.A., Baum, A., Fisher, J.D., Green, T.E., and Greene. T., *Environmental Psychology*, International Thomson Publishing, London, 1990.

Bechtel, R.B. and Churchman, A., *Handbook of Environmental Psychology*, John Wiley & Sons, New York, 2002.

Cromley, R.G., *Digital Cartography*, Prentice Hall, Englewood Cliffs, New Jersey, 1992.

Igarashi, Y. and Sadahiro, Y., A GIS-based system for visualizing regional images, paper presented at The 99th Annual meeting of the Association of American Geographers, New Orleans, March 2003.

Igarashi, Y., Visualization of regional image using GIS, master's thesis, Department of Urban Engineering, University of Tokyo, Tokyo, Japan, 2003.

Johnson, R.A. and Wichern, D.W., *Applied Multivariate Statistical Analysis*, Prentice Hall, Englewood Cliffs, New Jersey, 2002.

Keates, J.S., *Understanding Maps*, Addison Wesley, Edinburgh Gate, Essex, 1982.

Lynch, K., *The Image of a City*, MIT Press, Massachusetts, 1960.

MacEachren, A.M., *How Maps Work: Representation, Visualization, and Design*, Guilford Press, New York, 1995.

MacEachren, A.M. and Taylor, D.R.F., *Visualization in Modern Cartography*, Pergamon Press, Oxford, 1994.

Nielson, G.M., Hagen, H., and Mueller, H., *Scientific Visualization: Overviews, Methodologies, and Technologies*, Los Alamitos, IEEE Computer Society, New York, 1997.

Pia, *Gourmet Pia*, http://g.pia.co.jp/, 2003.

Slocum, T.A., *Thematic Cartography and Visualization*, Prentice Hall, Englewood Cliffs, New Jersey, 1998.

References

Bailey, T.C. and Gatrell, A.C., *Interactive Spatial Data Analysis*, Longman, Burnt Mill, 1995.

Burrough, P.A., *Principles of Geographical Information Systems for Earth Resources Assessment*, Clarendon Press, Oxford, 1986.

Cressei, N., *Statistics for Spatial Data*, John Wiley & Sons, New York, 1991.

Cronley, R.G., *Digital Cartography*, Prentice Hall, Englewood Cliffs, New Jersey, 1992.

Langran, G., *Time in Geographic Information Systems*, Taylor & Francis, London, 1993.

Laurini, R. and Thompson, D., *Fundamentals of Spatial Information Systems*, Academic Press, London, 1992.

MacEachren, A.M. and Taylor, D.R.F., Eds., *Visualization in Modern Cartography*, Elsevier, New York, 1994.

Okabe, A., Boots, B., Sugihara, K., and Chiu, S.N., *Spatial Tessellations: Concepts and Applications of Voronoi Diagrams*, John Wiley, Chichester, 2000.

Okunuki, K., Itoh, S., Okabe, A., Goto, Y., Kaneko, T., Shinoaki, S., Akita, Y., Kotsubo, H., Okita, Y., Tatematsu, T., and Shiozaki, G., (2001) A mobile GIS for fieldwork, *Abstr. Symp. Asia GIS*, 20 –21, 2001.

Reply, B.D., *Spatial Statistics*, John Wiley, New York, 1981.

Star, J. and Estes, J., *Geographic Information Systems: An Introduction*, Prentice Hall, Englewood Cliffs, New Jersey, 1990.

Upton, G. and Fingleton, B., *Spatial Data Analysis by Example*, John Wiley, Chichester, 1985.

Index